# SCIENCE TEACHING IN SCHOOLS

*Published by*
**Saurabh Publishing House**

# SCIENCE TEACHING IN SCHOOLS

**Bharat Singh**

Saurabh

Publishing House

4263/3, Ansari Road,
Darya Ganj, New Delhi-110002

**Published by :**
**SAURABH PUBLISHING HOUSE**

**Distributed by :**
**LOTUS PRESS**
4263/3, Ansari Road, Darya Ganj, New Delhi-110002
Ph : 32903912, 23280047 • E-Mail : lotus_press@sify.com

Saurabh Publishing is an imprint of
**Lotus Press**

**Science Teaching in Schools**


**ISBN** : 878–81–89005–51–1 (H/B)



Printed in India by Anand SONS, Delhi

# Preface

Teaching of Science requires a scientific attitude and a perfect method. Science education is the field concerned with sharing science content and process with individuals, not traditionally considered part of the scientific community. The target individuals may be children, college students or adults, within the general public. The field of science teaching comprises science content, some social science and some teaching pedagogy. The standards for science teaching provide expectations for the development of understanding for students, through the entire course of their education. The traditional subjects included in the curriculum are physical, life, earth and space sciences. In schools, the science teacher is responsible for teaching and preparing the practical equipment and making up the solutions, used in school science labs. His role also includes instructing and enriching students with practical skills, including class demonstrations for advanced techniques, across all disciplines.

Science education is the science teaching and learning that occurs within school curriculum in classroom and laboratory. This comprehensive and compact book fills a vacuum for a perfect reader-friendly book to help science teachers in teaching comfortably and successfully. This book, not so bulky, would expectedly prove to be an asset, as a reference tool for all teachers, students and general readers. We invite our readers to send their enlightening remarks for encouraging our future endeavours.

— ***Author***

# Contents

*Preface* v

1. **Introduction** 1

Right Way of Teaching • Estimation of Learning • Topics of Teaching • Teaching Whom? • Way of Teaching • Taking on Challenges • The Trainers • The Pupils • The Guardians • Institutional Authority • School Administrators and Education Officials • Present Condition • Analytical Attitude • Right Affair • Answerable Questions • Science Syllabus: an Urge • Science Subjects for New Age

2. **Science as a School Subject** 29

Science: A Need of Today's World • Everyday's and Everybody's Science • Science as a Subject • Fundamental Characteristic of Science • Outcomes of Learning Science • Science and Society

3. **Techniques of Teaching Science** 43

Teaching Approach: Selection • Trying-out Teaching Skills • Microteaching • Approach to Study • Task Method • Working Science • Process Mode • Environmental Mode • Utilisation of Various Methods • Scientific

Mode • Mode of Project • Doctrine for Choosing Teaching Techniques • Several Techniques • Lecture Method • Lecture Mode in Advanced State • Presentation Mode • Indefinite Method • Lecture Presentation Mode • Lecture Presentation Approach in Advanced State • Laboratory Methodology • Enquiry Project Methodology • Solution Giving Mode • Personal Guidance Mode • Modular Method • Student and Teacher Centred Method • Heuristic Approach • Enquiry Method • Debating and Clarification Mode • Programmed Guidance Approach • Method of Multimedia Package • Teachers and Science Education • Preparations by a Good Teacher • Tasks of Science Teacher

4. **Science Teaching Objects** 97

Teaching Objectives • Science Learning • Systematic Process • Awakening of Natural Interest • Scientific Attitude • Teaching Goals and Purposes • Educational Aims: Classification • Classified Aims of Cognition • Influenced Aims Classified • Influenced and Cognitive Domains • The Contrast • The Objectives • State-wise Programmes

5. **Organised Teaching** 135

Origin and Development • Hindrance (Terminal Objective) • Activity Interpretation • Entry Attitudes • Purpose of Attitude • Weighing Instruments • Instructional Series • Suitable Means • Programme Evaluation • Some Facts about the Unit • More About Baboon Frame • Typology of Programmed Units • General Aspects • Psychology Associated with Behaviour • Significances of Programmed Guidance

• Combining Practicals with Theory • Additional Programming • Flourishing a Programmed Guidance Unit

6. **Significance of Audio-Visual Devices** 159

The Demonstrations • Small Screen and Moving Pictures • Pictures and Radio Recordings • Significance of Slides • Slides' Effective Utilisation • Recordings and Radio • Sighted Symbols • Oral Symbols • Competent Utilisation of Audio-Visual Supports • The Significance • Experiences' Edger Dale Cone • Straight Determined Experiences • Strategic Experiences • Exaggerated Involvement • The Exhibition • Ground Visits or Expedition

7. **Curriculum Development** 191

Modern Scientific Development • Adequate Provision in Curriculum • Good Outcomes of Science Learning • Syllabus Formation: Abilities and Interests • General Education's Necessity in Science • Disciplinary Approach • Basic Course • Advanced Course • Concept and Theory

8. **Planning the Lessons** 273

Classroom Lesson Planning • Content of Lesson-plan • Challenging Learning • Stages Involved in Lesson Layout • Unit Layout • Planning the Lessons • Skill for Asking

*Bibliography* 297

*Index* 301

# 1

# Introduction

What do we teach in Science? Why should we teach what we teach in Science? How should we teach Science? Why should we teach the way we teach Science? How should we evaluate that students have understood what we taught them. These are some very important points to be considered by those who are concerned with science teaching. They may be science teachers, educators, researchers and curriculum developers. This is what is usually taught in Science Method Courses.

## Right Way of Teaching

There are various methods of teaching science "When a particular method is to be used," is very important.

In science there are some concepts and some skills "Temperature" is a concept, and "Measuring temperature or reading a thermometer," is a skill. 'Weight" is a concept and "Weighing by a spring balance, weighing machine, physical balance or chemical balance, ' is a skill. "Photosynthesis" is a concept, and "setting up and performing photosynthesis experiment," is a skill.

By researchers it has been found that concepts are best learned by Child Centred Approach (CCA), and skills are best learned by Teacher Centred Approach (TCA).

If you want to develop "creativity ' and "scientific temper," do not give them readymade answers, use "Scientific Method" or "Problem Solving Method". Similarly you can think of when a particular method of teaching science is to be used.

## Estimation of Learning

### *During Teaching*

When teaching science in a classroom, we ask some questions before we start the lesson. The answers of these questions give us some idea, whether the students have the previous knowledge or entry behaviours for the lesson. We ask some questions during the lesson and also encourage students to ask questions if they do not understand anything. Answers to such questions give us the idea whether the students are understanding what we are teaching them. We also ask some questions after the lesson, just to know whether students understand what we taught them in the lesson.

### *In Examination*

Let us look at our examination system. If we analyse the examination questions very critically, we find mostly knowledge questions are asked and very little emphasis is given to comprehension, application and skill questions. Our examination questions should test all the objectives — cognitive, affective as well as psycho-motor. Clearly defined objectives help us in effective and objective evaluation. Therefore, we should identify our objectives before-hand and state them very clearly. If our objective is just knowledge, we should not ask comprehension, application or skill questions; and if our objectives are knowledge, comprehension, application and skill, we should test all the objectives and not just the knowledge as it is usually the case.

If our objective is 'to develop the skill of reading a thermometer, our test should be, 'Give a thermometer to the student and ask him to read, instead of how will you read a thermometer? Explain with the help of a diagram. Our education is examination-oriented. Though we identify several objectives but they are just on paper. We try to achieve only those objectives which are usually tested in examination. So our examination system has a big hand of providing quality science education to our children.

## Topics of Teaching

What we teach in science is 'content'. Now the question arises, *'what is science'*? Science is a way of describing and explaining some aspects of the world around us. To the extent that a lot of human effort has already been expended in developing such explanations, our students do not have to 'start from scratch'. A large and ever-increasing body of scientific knowledge already exists, and evidently part of the task as teacher is to pass on some of it to the students.

But this is only part of the job, for science is also a package of *processes* by which we can increase our knowledge of external world. Teaching science implies involving our students in investigation, so that they become 'scientists for the day'. There are facts, theories, concepts and also a way of working, which together constitute 'the subject science.' Thus, science is not just content. Science is content plus processes. This is dual nature of science.

***Content***: One of the things you will be doing as a science teacher is familiarising your pupils with theories, helping them to develop concepts and appreciate the structure which connects them. However, so extensive and powerful are these networks of concepts that there is a tendency never to put pupils in the position of scientists. Thus, pupils come to believe that science provides no opportunity for the expression of differences of opinion, for imagination and for creative work. They feel that there is no scope for personal contribution, that knowledge is embodied in the natural world waiting to be discovered and, once discovered, must be accepted without question. But scientist makes his own observations and deduces generalisations from them.

Content to be taught at a particular level should be carefully selected. It is a hard task. Scientific knowledge keeps on increasing. It doubles every decade. Our students are to keep pace with increasing scientific knowledge. We should teach them what they really need at a particular level. If it is more to be completed in a classroom in the allotted time, we should develop some techniques so that students could learn a part of content on their own outside the classroom.

***Presentation of Content***

'What we teach' should be compatible with:

- cognitive level of students,
- identified objectives, and
- existing classroom conditions.

The 'content' presented to the student should be such that he could understand it, and not just memorise. Learning is understanding, and not memorising. In every class there is some content in science, which most of the students memorise without understanding. Such content should not be included at that level, but shifted to a level where students can really understand it.

The content in the textbook should be presented in such a way that pupils like it. They take interest in it. Sometimes the content given in the syllabus is compatible to the cognitive level of pupils, but it is presented in the textbook in such a way that the pupils have difficulty to follow it. Therefore, they do not like to read such textbooks. According to Professor Jack Carter of University of Colorado, USA and Director of B.S.C.S. Project (AISTA Silver Jubilee Conference, New Delhi, December 1981), "the content presented in our science textbooks is just like a scientific paper which is meant for those who are experts in science, rather than for those who are students in science."

Syllabi given by the education departments or education boards are usually outlines. Therefore, different authors go to different depths for the same content when writing the textbooks. If our science syllabi are well defined in the form of 'objectives written in behavioural terms' the authors will know exactly what they are to write, students will know what they are to learn and teachers will know what they are to teach, and paper setters will know what is to be evaluated. All of them will be on the same track, and will not go astray.

### *Science Processes*

The current innovations in schools emphasise the processes of science, the ways in which scientists advance their knowledge and solve problems. Science should be presented to pupils as a way in which they can conduct an enquiry into the nature of things as well as a body of information built by other people. In the past, the processes have been neglected, and school science has been concerned almost exclusively with the 'body of information, the concepts, principles and techniques developed by scientists; too rarely have pupils been made 'scientists for the day,' encouraged to become personally involved in solving problems, in discovering some science for themselves.

The teachers' guides of most recently developed school science programmes in United States and United Kingdom suggest ways in which pupils can become involved in scientific enquiry. They ask to follow the stages: Problem-Hypotheses — Experiment — Result, in solving problems. If you as a science teacher, are to engage your pupils in the processes of science, it is important that you should have thought carefully about what these processes are. The sequence of development of the thirteen science processes is as follows:

1. Observing,
2. Classifying,
3. Using numbers,
4. Measuring,
5. Using space-time relationships,
6. Communicating,

7. Predicting,
8. Infering,
9. Defining operationally,
10. Formulating hypotheses,
11. Interpreting data,
12. Controlling variables, and
13. Experimenting.

### *Nature of Science*

Science is a broad based human enterprise that can be defined differently from different viewpoints. The layman might define science as a body of scientific information, the scientist might view it as a method by which hypotheses are tested, a philosopher might regard science as a way of questioning the truthfulness of what we know. All these views are valid, but each presents only a partial definition of science, only collectively do they begin to define the comprehensive nature of science. Therefore, science should be viewed as a way of thinking in the persuit of understanding nature as a way of investigating claims about phenomena and as a body of knowledge, that has resulted from enquiry.

### *Science as a Way of Thinking*

Science is a human activity that can be characterised by the thinking that occurs in the minds of people who participate in it. The mental work of scientists illustrates human kind's curiosity and desire to understand phenomena. These individuals possess attitudes, beliefs, and values that motivate them to answer questions and solve problems. Scientists are driven by enormous curiosity imagination and reasoning in their quest to figure out and explain natural phenomena.

Their work, as viewed by many philosophers of science and cognitive psychologists, is a creative activity whereby ideas and explanations are constructed. Therefore, the thinking and reasoning of scientists as they go about their work offer important causes regarding the nature of science belief. The tendency for scientists to find out seems to be motivated by their belief that a set of laws of nature can be constructed from observation and explained by thought and reasoning.

### *Science as a Way of Investigation*

Science as a way of investigation illustrates many approaches to constructing knowledge. Science has many methods which demonstrate human kind's investigation for seeking solutions to problems. Some of the approaches used by scientists rely heavily on observation and prediction as in astronomy and ecology other approaches rely on laboratory experiments that focus on cause and effect relationships such as those used in microbiology and other sciences. Among the many processes associated with science and enquiry as discussed above are observing, classifying, using numbers, measuring, using space-time relationships, communicating, predicting defining operationally, inferring, interpreting data, controlling variables, and experimenting.

### *Science as a Body of Knowledge (Product)*

The body of knowledge produced from the scientific disciplines represents the creative products of human invention that have occurred over the centuries. The enormous collection of ideas pertaining to the natural and physical world is organised into astronomy, biology, chemistry, geology, physics and so on. The result is a compilation of carefully catalogued information containing many types of knowledge, each of which makes its own unique contribution to science. The facts, concepts, principles, laws, hypotheses, theories, and models from the content of science. These ideas poses their own specific meaning, which cannot be understood apart from the processes of enquiry that produced them. The facts of science serve as the foundation for concepts, principles, and theories.

According to *'The Columbia Encyclopaedia'* (1963), "Science is an accumulated and systematised learning, in general uses restricted to natural phenomena. The progress of science is marked not only by an accumulation of facts, but by the emergence of scientific method and of the scientific attitude".

From the above definition three basic principles of the nature of science can be identified (1) an accumulated and systematised body of knowledge, (2) the scientific method of investigation, and (3) the scientific attitudes or ways of thinking. The first point indicates the product of science, while second and the third points indicate the process of science. In other words, Science is both a product and process.

Thus, Science has a dual nature.

1. Content of science. 2. Process of science.

## Teaching Whom?

It can be discussed into two parts:

1. Why should we teach science?
2. Why should we teach what we teach in science at a particular class level?

### *Why should we Teach Science?*

It is very unfortunate that we as science educators never practically think at 'why.' The principal goal of science education is to create men who are capable of doing new things, not simply of repeating what other generations have done — men who are creative, inventive and discoverers. If that is why we teach science, then how far our science education is compatible to achieve this goal?

The key to national prosperity lies in the effective combination of three factors — technology, raw materials and capital, of which perhaps the first one is the most important. Since the creation and adoption of new scientific techniques, technology can, in fact, make up a deficiency in natural resources and reduce the demands on capital. But technology can only grow out of the study of science and its applications. Technology serves man. It gives individuals and nations roots, a frame of reference, horizons, a world view and inner freedom.

Science and technology are used to promote the establishment and furtherance of the community of man to the community of nations. We teach or learn science for a variety of reasons.

1. Science is fundamentally concerned with exploring and interpreting the physical world through the three fundamental areas of Physics, Chemistry and Biology. Physicists are concerned with the exploration of energy and general properties of inanimate materials, chemists with the particular properties of inanimate materials, and biologists explore the behaviour and properties of animate materials. The physical and biological world is of fundamental human interest and man has a basic motivation to understand and control the physical world in which he has his being.
2. We live in a scientific and technological age and no citizen can function effectively in a developed society without a basic scientific literacy and certain elementary skills. Every citizen needs to live a healthy life with proper sanitation and clean surroundings. The knowledge and skills required by a modern householder in dealing with electrical, plumbing, and human first-aid may be cited as some justification for teaching science to all children in school.
3. We depend upon scientific knowledge and understanding for economic and material advancement. Science has provided so many aids for 'the good life' from bicycle to jet aircraft, antibiotics to heart surgery, radar to colour television, and fertilizers to plant growth harmones. Although we are beginning to realise many follies in the large-scale application of science and technology in terms of a net environmental damage, the fact is that we will need science again to rectify these follies (like application of science and technology in a better way which would not affect the environment adversely) and to enable more people to live on this planet in conditions which do

greater justice to the dignity of man. The priority which the developing countries like India are giving to basic science education for public health and agriculture can clearly be seen.

4. Science cannot be used in society without a body of men and women who have been specially trained for science-based vocations in industry, research and teaching. A vocational justification is, therefore, possible for science teaching.
5. Science, if studied properly, develops power of thinking, reasoning, curiosity, open-mindedness, and ultimately develops scientific attitude or scientific temper. Thus, science could be aimed as an instrument for social change towards a better society.
6. Professor Yashpal (1992) is of the opinion that, instead of considering science as an extraneous activity or as a tool for providing the means to a good life, usually borrowed from outside, we should treat it as a part of the culture of society, integral to our living and thinking and connected to the deepest questions we ask in regard to who we are and where we come from. Although, the tools that technology provides in the form of increased means of productive goods and services, means of communication and transport are important, but if we make science as a part of our culture, they would all come, much more naturally, more creatively and more in tune with our living.
7. Science is essential for helping us to comprehend and establish connections with the physical and biological world around us. It familiarises students with the empirical modes of investigation. Understanding its conceptual structures and methodologies are an important part of an educated person. It can help in the development of intellectual skills and motor abilities like comprehension and application of understanding to new contexts and

accurate observation of phenomena. Scientific knowledge not only influences our thinking about human, political and social affairs but also affects our thinking about ourselves. Its importance for arousing scientific curiosity and in developing scientific attitudes, as well as educating people for science based vocations in industry, research and training cannot be overemphasised. The scientifically literate person has a knowledge base of facts, concepts, conceptual networks, and processes, which enable the individual to continue to learn throughout life.

### *Why should we Teach what we Teach in Science at a Particular Class Level?*

Who decides what is to be taught in science at a particular class level, i.e., science syllabus for various classes (I-V, VI-VIII, IX-X, XI-XII)? Perhaps those who never taught those classes. Science teachers who teach those classes are not empowered to make such decisions. Therefore, some topics become too difficult to be understood by the children. It will be advisable that syllabus framers should also involve concerned science teachers when they frame science syllabus for various classes. Syllabus framers should be answerable, why they put this science content in the syllabus. They should objectively say, "We put this science content in the syllabus because children need it, children will understand it and children will enjoy it."

## Way of Teaching

"To learn science is to do science, there is no other way of learning science," D.S. Kothari. This is how we should teach science. Do we actually do so?

Several methods of teaching science are taught in B.Ed. classes but what B.Ed. students mostly practise are 'lecture method' or lecture-cum-demonstration method' during their practice teaching. And so they usually use the same two methods when they become regular science teachers. They should practise all the methods which they learn in theory.

Then they will be able to use what best fits in the actual classroom conditions.

Our science courses are mostly content-oriented. When content worship is becoming old fashioned in developed countries, our country is still behind it. Whenever the courses are updated more content is added. Courses become heavier and heavier while time allotted in the school timetable remains just about the same. Thus, students are to cover much more content in comparatively much lesser time. Moreover classes are crowded and teachers are heavily loaded. Thus, in spite of sufficiently available science equipment for experiments and demonstrations, content is covered by asking students to read books, or giving lectures to cover more content in lesser time without demonstrations and experiments.

Under such classroom conditions we are to devise some self-learning techniques in teaching science. Programmed Instruction is one such device. When Self-learning Modules, Programmed Instruction Materials in science or Adjunct Programmes on various science textbooks are used, students will be able to cover a part of the syllabus on their own, and teachers will get some time for individualised instruction. If one were to list expressions that best describe science education for the twenty-first century 'Individualised Instruction' certainly has to be included. The Programmed Instruction and Modular Approach are some of the efficient ways of providing individualised instruction. If our science teachers could have well-sequenced programmed materials and self-learning modules, it would be great help to them since they are heavily loaded. Also if the programmed materials and self-learning modules provided activities of 20-25 minutes duration, students would have more opportunity to do the experiments themselves, even in smaller periods.

### *Constructivism: A Dominant Perspective about How Children Learn Science*

Constructivism is a general name given to the dominant perspective on learning and especially in science education.

This approach of teaching learning is quite different from traditional views.

"I hear and I forget, I see and I remember, I do and I understand", must be a familiar phrase for you, but this phrase shows the intent of constructivism. Constructivism is not entirely new. Contemporary researchers from various countries have updated the theories and methods and synthesised several dominant perspectives on learning. This perspective is evolved from the research and theories of Piaget, Vygotsky, Bruner and philosophy of John Dewey. There is a consensus among many philosophers of science, psychologists, science educators, scientists and people concerned with the children's learning that learners must construct and reconstruct their own meaning of ideas about how the world works.

Joseph Novak (1986) defines constructivism as the notion that humans construct or build meaning into their ideas and experiences as a result of an effort to understand or make sense of them. According to Duckworth (1989), all people have their own understanding, one cannot make them believe anything unless they construct it by themselves.

There is no one constructivist theory of learning. Therefore, students can be encouraged to learn by reinventing the wheel for themselves, so that the learner does the discovering by forming mental connections. The understanding or conceptual clarity is deeper and retention is greater. The teacher mediates the learning environment and thus plays an important role. The hands on activities in science, manipulative skills, process based learning share some common intentions with coustructivism, which is becoming popular in education. The constructivist teacher must fulfil many roles, helping the learner to think critically and use information creatively to construct scientific ideas, concepts and largely function as a facilitator of knowledge construction in order to help the learner how to learn. This conceptual clarity through construction of knowledge is ultimately reflects through their attitudes. This should be the constructivist role of a teacher.

## Taking on Challenges

Science Teaching today is not so simple as it was in early-fifties. But even in those days use of audio-visual aids was quite common, though students and parents were not as demanding as they are now. In those days there was not so much competition, there was not so much load on students, and therefore there was no need of rote memorisation. Now with the loaded curriculum students cannot escape the drudgery of mugging textbooks and learning without tears and fears.

Now in the existing school and classroom conditions Science Teaching is a Challenge, which a science teacher alone is to face, but he cannot do much unless he has support from the Principal, Education Officers, SCERTs, IASEs, School Education Boards, DIETs, Parents as well as NCERT, NIEPA and NCTE.

Some Teachers, Students, Parents, Principals, Education Officers, and school Managers were interviewed. What they feel about Today's Science Teaching — A Challenge, is given below.

## The Trainers

1. We should have 24-30 periods a week with no substitution periods so that we could get enough time for planning, evaluating copies, preparing demonstrations and experiments.
2. Well-equipped Science Labs, with enough audio-visual aids for demonstrations and experiments.
3. Lessons to be observed and necessary guidance be given for better science teaching.
4. Giving some demonstration lessons with up-to-date educational technology.
5. To be trained in the latest techniques of teaching science and in new science content enrichment (pre-service and in-service).

6. To be trained in continuous and comprehensive evaluation.
7. Qualified and well trained Lab. Assistants.

**The Pupils**

8. Good science teaching, every difficulty to be removed in the class, teachers should not proceed further unless all students understand, to avoid private tutions, so that we could have good marks in X and XII to get admission in higher studies.
9. Less homework, more work in school.
10. Less Course, useless course to be removed.
11. Two Science Courses in X — B-Course for those who will not take science in XI-XII; A-Course for those who will take science in XI-XII; A-Course should cover all Entry Behaviours for XI-XII Science (Physics, Chemistry, Biology).
12. Coordination in Theory and Practical, not as we have now — Theory in IX and Practical in XII.
13. Less Extra Classes — in Extra Classes only Remedial and Enrichment Work, Courses to be completed in regular classes.
14. Maximum and Effective use of Audio-Visual Aids, Hardwares and Softwares in Science Teaching, Science Teaching should be Activity Based — Demonstrations, Experiments, Field Trips, Investigatory Science Projects.
15. Use of computers to teach some portion of Physics, Chemistry and Biology.
16. Use of Self-Learning Multimedia Packages-printed, audio, video, audio-video, computers.
17. Make Science Teaching enjoyable at least less Boring.

**The Guardians**

18. Good Teaching - No Need of Private Tutions.
19. Remedial and Enrichment Extra Classes.

20. Excellent Results in X and XII to get admission in higher studies without outside coaching.

**Institutional Authority**

21. Intensive Training in Educational Administration, Planning and Management — Pre-service and In-service.
22. Needed support and guidance from officers of education departments.
23. Science Textbooks should be tried out and revised before implementation — Formative Evaluation Before Summative Evaluation.
24. Officers of education departments should also be given proper training to guide and supervise the work of Teachers and Principals, and give some concrete suggestions for improvement of science teaching.
25. Science supervisors of education departments to be very active for giving proper guidance to Science Teachers for quality science education and good science results.

**School Administrators and Education Officials**

1. All teachers should teach and all students should study.
2. 100 per cent results with all First Divisions and Distinctions.
3. Effective use of Audio-Visual Aids and Educational Technology.
4. Use of computers in Learning Science.
5. B.Ed. Courses should be restructured to enable science teachers to teach science very effectively in existing classroom and school conditions, one science teacher should be able to teach full science upto class X.
6. Timely Need-Based and Effective In-service Training of Science Teachers to enable the teachers to teach science very effectively to give good results.

All these issues are to be very seriously taken up by concerned people and agencies to face the challenge in Science Teaching.

## Present Condition

Let us examine the present situation of science education in our country. First of all let us have a look at the objectives of science teaching as stated by Prof. B. Sharan of NCERT, which are as follows:

1. To make aware of those aspects of science that are environmental based and life centred.
2. To emphasise the scientific method which has helped in solving problems.
3. To train students in the use and maintenance of science equipment.
4. To make students creative.
5. To develop the mental faculty of open-mindedness.
6. To prepare children for change of life.

Now you can yourself examine, how far our existing science curricula, teaching methods and evaluation techniques are compatible to achieve these objectives.

The curricula have been drawn up and textbooks written by those who perhaps have never taught in schools. They have tried to compress in the textbooks, a variety of topics some of which have little utility or relevance. In spite of the Ishwarbhai Patel Committee on Curricula Reforms for the ten-year school submitted in 1977, the curricula have not been updated. Even the Minister of Education presiding over the annual general meeting of the NCERT on March 11, 1981 regretted that attempts at improving the quality of education in terms of upgrading curricula were not heartening.

One of the recommendations of the Education Commission under the chairmanship of Prof. D.S. Kothari (1964-66) was that science should be made a compulsory subject in school education. The recommendation was accepted and science was made compulsory upto class X in several states and Union Territories. 10+2 Education Scheme started in July 1975. Science courses at 10+2 stage were framed in a hurry and they were implemented without any try-out.

When framing the courses, not many teachers who actually teach the students at this level were involved, and editorial boards were made of people from Universities and Institutes of Technology. Ishwarbhai Patel Committee for reviewing Class I-XII courses was asked to submit its report within two months' time. Therefore, *ad hoc* deletions were made. Sometimes diagrams were deleted while content related-to those diagrams remained.

In the old 3-years Higher Secondary Scheme (IX-XI) about one-third of students who were really interested in science used to take science in class IX. But now science is also a required course like other subjects upto class X. Therefore, majority of the students (about two-thirds) who perhaps do not have the interest and aptitude for science have to take it. Now the question arises, must every student be forced to learn a subject for which he has no aptitude or a special science course be developed for such students so that they are also attracted towards learning science.

The former Prime Minister Morarji Desai, while addressing the members of the Review Committee on the curriculum for the ten-year school at New Delhi stated, "The books that I did carry in College are being carried by School students today. The students are burdened by how many books, I do not know." Knowledge of science almost doubles every decade. We are to keep pace with this new development. Then the problem is how much knowledge in science should be given to a child at a particular level, so that he is not burdened.

The goals and expectations of the science education change with time and with them the curriculum. Updating, revising, reorganising, or adopting science curricula are not sufficient to reduce the present state of fragmentation and narrowness, what is necessary is the design of science programmes for the-non-specialist, the common man.

The social imperatives of education and specially science education have recently been attributed new importance and urgency. The long-term social problems of: (1) environmental pollution, (2) food shortages, (3) limitations on available energy, and (4) expanding population demand, change the curriculum emphasis and modes of teaching. Students are not only to be informed about these social problems, they also need practice in making decisions, so that as adults they will be wiser in individual and collective decision-making.

Since these social imperatives do not conform to the traditional academic boundaries of the sciences, or even of any subjects commonly taught in schools, new curricular patterns must be generated, developed, tested, and evaluated for evidence of effectiveness. Relevant research is needed not only in the development and testing of these new instructional materials, but more basic studies are required on the ways, means and amounts of various experiences which modify the attitudes and enhance decision-making skills.

It is time that the country's intelligentsia, educational planners, educational administrators, educationists, politicians, teachers and parents should give some serious thought to what direction the country's education system should go, and what should be the place of science in the curriculum. They owe it to their children and to the future generations. The time seems to be ripe for a serious national debate on the 'future of science education' taking into a serious view of its past and present.

We always say that we should teach science in such a way that we are able to develop scientific attitude and scientific temper among our students. This remains our objective of teaching science. How far we are able to achieve this objective? What is scientific attitude? What is scientific temper? Is there any relation between these two? Can scientific method help us to develop scientific attitude and Scientific temper among our students?

## Analytical Attitude

***Before Talking about Scientific Attitude Let us Discuss What is Attitude ?***

Attitudes are mental state or individual outlook towards people, objects, events, etc. Attitudes are developed and learned/ they are not inborn. Attitudes can be changed through experiences especially in children. They are the result of experiences.

Scientific attitude refers to an individual's outlook towards life. It means a willingness to adopt scientific approaches and procedures for resolving issues/ assessing ideas or information. Robert Ebel (1997)/ defines attitude as a mental condition/ a stabilised mental set which expresses itself in a tendency to react to any member of a class of stimuli in the same general way. Scientific attitude predisposes a person to engage in responsible action after weighing the possible consequences of alternative options/using rational arguments based on evidence. She distinguishes between scientific evidence and personal opinion and between reliable and unreliable information, remains open to new evidence and the tentativeness of scientific knowledge and recognises that science and technology are human endeavours.

Some other manifestations of the scientific attitude are/ questioning/suspension of judgement, appraisal of different sides of an issue, verification of knowledge, curiosity and appreciation of the natural and human-made world, open mindedness/intellectual honestly/cooperation and tolerance of the point of view of other people.

The complex interactions of science with society need to be understood by individuals and applied in everyday decision-making. For improving the quality of life through economic and material development/ we depend on the knowledge of science. Developments in science have prepared the way for a more effective/ practical mastery of the environment and the technological innovations in agriculture, industry and medicine have produced transformations in traditional patterns of social living.

In the words of Indira Gandhi (1982). "Science has not only radically altered man's material environment/ but what is of still deeper significance/ it has provided new tools of thought and has extended man's mental horizons. It has thus influenced even the basic values of life/ and has given to civilization a new vitality and dynamism".

***Scientific attitude or Scientific temper***

1. open-mindedness;
2. objective decision-making;
3. critical thinking;
4. not believing what other people say unless convinced what is being said is correct;
5. desire for accurate knowledge;
6. confidence in procedures in seeking knowledge;
7. expectation that the solution of the problem will come through the use of verified knowledge;
8. using scientific method (problem, hypotheses, experiment, conclusion) in decision-making.

*Recognising a person having scientific attitude (Scientific Temper) through the following episode in which there are four main characters — principal science consultant physics teacher and science students.*

## Right Affair

A Science Consultant went to a school. As soon as he entered the Principal's office, Principal requested him to go to Ms. X's Physics class as she does not do anything and we have lot of complaints against her. The Science Consultant went to her class and observed her lesson. Here are his remarks "The teacher was teaching waves by lecture-demonstration method. She was demonstrating longitudinal and transverse waves, crests and troughs, rarefaction and condensation, nodes and antinodes, frequency and wave length with the help of a SLINKY. It was a very lively class.

Students were really enjoying and participating. There was very healthy teacher learner interaction in the class. Questions between and after the lesson were thought-provoking. Homework copies and practical notebooks were checked up to date.

The physics lab was well set for the day practical work. "The Principal was shocked to listen to these comments. She spoke to the Science Consultant, "Are you sure, these are your comments." "Yes, these are my comments. She is an outstanding physics teacher. How did you know that she does not do anything? Who was complaining against her, the science consultant asked. Some teachers were telling, the Principal replied "Do not believe what others are saying, unless you are convinced that what they are saying is correct," the Science Consultant advised the Principal.

## Answerable Questions

1. Who has scientific temper — Principal or Science Consultant ? Justify your answer
2. Do all science teachers have scientific temper?
3. Do you have scientific temper?
4. Do non-science teachers also have scientific temper?
5. Do science or non-science principals have scientific temper?
6. Do scientists have scientific temper?
7. Do non-scientists have scientific temper?
8. What is scientific temper?
9. How will you recognise whether a certain person has scientific temper?

10. (i) What is the difference between "scientific attitude" and scientific temper ?

(ii) What are the characteristics of a person having

(a) scientific attitude ?

(b) scientific temper?

(iii) A Science teacher has the following characteristics:

(a) She is open-minded.

(b) She has a burning desire for the acquisition of correct knowledge and search for truth.

(c) She has confidence in her abilities to seek knowledge with her own efforts.

(d) She has an adequate ability of problem-solving and believes that the problems can be solved through proper efforts involving scientific observations and experimentation.

(e) She has spirit of curiosity.

(f) She believes in cause and effect relationship.

(g) She believes in the theory of evidence.

(h) She has love for truth.

(i) She adopts scientific method (problem, hypotheses, experiment, conclusion) in her thinking and working.

(j) She teaches science by scientific method, and uses the method in her decision-making.

*Now answer the following questions.*

1. Does she have (a) scientific attitude (b) scientific temper?
2. Which of the above characteristics show that she has: (a) scientific attitude, (b) scientific temper?
3. Do you find any difference between scientific attitude and scientific temper?
4. How to develop scientific attitude or scientific temper?

***Try for the following:***

1. satisfaction of curiosity;
2. to get rid of superstitions;
3. proper way of teaching with no ready-made answers;
4. utilising co-curricular activities like working on investigatory science projects;
5. open classroom environment;
6. inspiring for the study of scientific literature;
7. encouraging day to day problems by scientific method;
8. develop in students well defined abilities such as the: (a) spirit of enquiry, (b) creativity, (c) objectivity, (d) courage to question, (e) problem solving and decision-making skills;
9. self example by the teacher.

## Science Syllabus: an Urge

Now Science is for all upto Class X. Then what type of science curriculum our children need to be functionally literate, as well as achieve optimally in science. Several children, parents, teachers, principals and administrators were interviewed. Here are some suggestions for restructuring school science curriculum at all levels.

1. For children to do well in science, we need to identify the basics in science. Once the basics have been chosen and implemented, children will have key facts, concepts, and generalisations to achieve. Teaching science will be more professional once the basics have been identified. A sower (the science teacher) that goes out to sow will then reap not ten nor twenty fold, but hundred fold. Why so? Time spent in teaching science will be spent on the relevant basics, and not upon the irrelevant.
2. Basics are to be stated in measurable terms. There is no reason for keeping the children in the dark as to what they are to learn. Measurably stated objectives, announced

clearly to children, will assist them to achieve as much as possible. Children achieve poorly in science if the evaluation techniques are not matched with the precise objectives.

3. Science instruction should be based on the interests of children. Good science teachers find out what children are interested in, and bring these interests into each lesson. Good training (pre-service and in-service) makes good science teachers. Good science teachers make good science students. If these good science students are permitted to give their input into the science curriculum, they will bring many many items of science to make science interesting.
4. Research data indicates how important parents are in helping their children learn. If parents assist their children in learning, achievement continues to rise. Teachers, administrators and parents should learn to work cooperatively to increase children's achievement in science. There would be some parents whose input might be taken into the science curriculum.
5. We need to stimulate children to identify and solve problems within the science curriculum. Subject matter may become outdated due to new research findings in science, but problem solving is always to stay. We should not focus so much on the content of science, but on processes of science.
6. In the past the slow learners received an inferior science education. Science teachers should be trained to identify slow learners and teach them science in such a way that they also enjoy science like normal children.

## Science Subjects for New Age

Now we entered twenty-first century. What types of Teaching of Science Courses we are going to have? It all depends upon what type of "Science Education" we will have in Twenty-first century. Let us study the existing Teaching of Science

Courses, so that they could accordingly be revised for the Twenty-first Century.

### *Teaching of Science Courses for Universities*

At present there are different Teaching of Science Courses, in various universities of the country. As a sample few of them are given here for study.

1. Syllabus Outline for Science Method Course;
2. Syllabus Outline for Integrated Science Method Course;
3. Syllabus Outline for Secondary Physics Method Course;
4. Syllabus Outline for Senior Secondary Physics Method Course;
5. Syllabus Outline for Secondary Chemistry Method Course;
6. Syllabus Outline for Senior Secondary Chemistry Method Course;
7. Syllabus Outline for Secondary Biology Method Course;
8. Syllabus Outline for Senior Secondary Biology Method Course;
9. B.Ed. (4 years) Science Course

#### *Syllabus Outline for Science Method Course*

Paper — 1. Natural Science

Paper — 2. Pedagogy of Environmental Studies

Paper — 3. Pedagogy of Natural Science

#### *Syllabus Outline for Science Content Course*

Paper — 4. Physics I

Paper — 5. Physics II

Paper — 6. Chemistry I

Paper — 7. Chemistry II

Paper — 8. Biology I

Paper — 9. Biology II

### *Teaching of Science Course for DIETs*

In some universities Teaching of EVS (science) course is also introduced. One District Institute of Education and Training (DIET) is to be started in each district of the country. So far (1997) 425 DIETs have been established. These are coming up since 1988-89. NCERT has developed 2 years Elementary Teacher Education Diploma Course for Pre-service Teacher Education (PSTE) students of the DIETs. In 1st year they are trained to teach primary classes (I-V). NCERT developed DIETs syllabus includes a good Teaching of EVS (science) course.

This may be adopted or adapted by universities if they introduce this course at B.Ed. level. In 2nd year DIETs' PSTE students are trained to teach upper primary or middle classes (VI-VIII). NCERT syllabus for DIETs also includes a good Teaching of Integrated Science Course. This course may be compared with "Syllabus Outiine for Integrated Science Method Course"and adopted or adapted by universities when they introduce this course at B.Ed. level.

# 2

# Science as a School Subject

The study of science, as a subject is essential, nowadays. In fact, a citizen of a modern world sees the countless manifestations of science all around him. There is no aspect of man's life today which has not been influenced by science one way or the other. This is because we are living in an age of scientific culture. Science has shrunk the world and totally changed the human outlook. In fact, science now has an all-pervading influence on every sphere of human activity. Further, modern science is no longer confined to the surface of this globe; its sphere of achievements reaches beyond the earth. Today, science is not influencing our planet alone. It is affecting the whole universe.

## Science: A Need of Today's World

Science is the key word today. The modern civilization is a scientific civilization. This is an age where the modern society is completely drawn into the scientific environment; and science has become an integral part of our life and living. Now, we cannot think of a world without science. The wonderful achievements of science have glorified the modern world.

Science, as a subject is evergrowing. In recent times, there has been rapid addition of knowledge to the world of science. Great advancements of science and technology and the use of these scientific achievements in promoting the well-being of mankind through their application in the field of industry, communication, transport, engineering, agriculture, medicine has made science more important than ever before. Science has, in fact, radically transformed the material environment of the citizens of the modern world.

In addition to the immense and far-reaching material benefits of science, as the Indian Education Commission puts it "... but even more profound is its contribution to culture. Science is liberating and enriching the mind and enlarging the human spirit. Its fundamental characteristic has turned out to be the possibility of unlimited growth. Every advance in science deepens our understanding of nature but it also heightens the sense of ignorance.

Nature is inexhaustibly knowable. Nothing comparable to the scientific revolution in its impact on man's development and outlook has happened since the Neolithic times." We have to live in a scientific civilization whether we like it or not. Science is no longer confined to a few seriously devoted persons. Since living in the present world invariably warrants, to variable degrees, knowledge of scientific facts and laws, science has now become everyday science for everybody.

## Everyday's and Everybody's Science

Teaching of everyday science for everybody has become an unavoidable part of general education. Nobody questions its inclusion as a subject in the school curriculum. It is included in a school's curriculum for the same reasons as any other subject, but in addition science inculcates certain special values peculiar to it and which no other subject can provide. It is a part of liberal education. But besides satisfying the usual needs for its inclusion as a subject in the curriculum—such as intellectual, cultural, moral, aesthetic, utilitarian as well as vocational values—science learning provides training in scientific method and also helps to develop a scientific attitude of mind in the learner.

The qualities imbibed by the learner through learning science are valuable for a citizen living in the society. Science is now a compulsory subject in every system of school education right from the elementary stage. In the past science had to struggle long and hard for its rightful place in the school curriculum. There was a time when science was considered an inferior subject to study and the meritorious students were supposed to study classics and mathematical subjects.

New ideas or inventions in science were not immediately accepted in the society and were looked upon with suspicion. Any new idea that went against the prevailing beliefs and codes of the time was condemned. Even Galilio Galilei, after the invention of the telescope had to write to his friend, the German astronomer Kepler, "My dear Kepler, what would you say of the learned here, who replete with the pertinacity of the asp, have steadfastly refused to cast a glance through the telescope? Shall we laugh or shall we cry?"

There was a time when many scientists refrained from making their new scientific ideas public for fear of social repercussion. Experimentation or practical work in science was also considered an inferior job. The real significance of science was never realised till the beginning of the nineteenth century and that is why we see that science had no place in the school curriculum during the earlier days. The pursuit of science was either a hobby of the people with means and leisure or the solitary efforts of someone with scientific talents. Even up to the middle of the nineteenth century, when sporadic efforts were made to introduce the study of natural science at the universities, it did not attract many students.

## Science as a Subject

Science was never a course of regular study either in universities or in schools. Since the beginning of the eighteenth century, however, a number of societies were formed in Europe and individual efforts were made for the popularisation of science. Such efforts have a long history beginning with Roger

Becon of the thirteenth century to Francis Bacon, Gilbert, Herbert Spencer, Tyndall, Huxley, Faraday, Armstrong and others who made efforts to popularise science. They pointed out its necessity for an all-round education of man. Benjamin Franklin did the same in America. Their efforts resulted in creating favourable opinions in society towards securing the rightful place for science as a subject to be studied.

Herber Spencer not only advocated the teaching of science but also proved science to be the most worthy subject of study from the utilitarian point of view in his book *What Knowledge is of most Worth.* On the other hand, T.H. Huxley wanted to show in his essays that science was beautiful. He pleaded for the inclusion of science in general education emphasising that learning of science can develop various favourable outlooks in the learner. It needed great effort and hundreds of years to popularise science even in its homeland, that is, Western Europe where modern science originated. The proceedings of the Royal Institute, London, record that the scientist, Michael Faraday, urged thus in 1858: "As a branch of learning, men are beginning to recognise the right of science to its own particular place; but now the fitness of university degrees in science is under consideration, and many are taking a high view of it, as distinguished from literature, and think that it may well be studied for its own sake, i.e., a proper exercise of human intelligence, able to bring into action and development all powers of the mind."

In India, the teaching of science followed the developments in England. But whatever development took place in India in the teaching of science was extremely slow. The teaching of science in schools in India was conspicuously deplorable till after 1953 when the Secondary Education Commission, popularly known as the Mudaliar Commission, recommended teaching of general science as a compulsory subject at the secondary school stage Galilio Galilei once said that in questions of science "the authority of a thousand is not worth the humble reasoning of a single individual." Though he referred to the faculty of theoretical reasoning, he was the first experimental scientist. While learning science, the learner develops certain faculties through reasoning and experimentation which no other subject can provide.

Science, as a teaching subject, possesses the various values as any other subject for which it is included in the curriculum as a teaching subject such as intellectual values, utilitarian values, cultural values, moral values, and aesthetic values. But in addition, the study of science inculcates certain other disciplinary values peculiar to it which cannot be provided through other subjects. Considering the subject from the intellectual point of view, science is the most inexhaustible storehouse of knowledge. It is opening new horizons of knowledge everyday and is continuously adding to the cultural heritage of mankind the ever-increasing new knowledge, new explorations, and new ideas. Along with acquisition of expanding knowledge, it makes man aware of the vastness of the unknowable in the universe.

## Fundamental Characteristic of Science

The fundamental characteristic of science is its possibility of unlimited growth. Since Nature is an inexhaustible source of knowledge, science as a subject, offers the widest range of knowledge to the learners. As such there is infinite scope for broadening one's horizon of knowledge. The world-wide researches and experimentations in the field of science and the advancements in the area are adding everyday to the already accumulated enormity of scientific knowledge. Acquiring knowledge for the sake of knowledge itself is one of the characteristics of civilization. In fact, science is both the foundation as well as structure of our modern civilization. It has exposed to mankind infinite avenues of knowledge in nature, living and non-living—the world we perceive and also the world beyond human perception thereby makes us conscious of the unknown to be explored. These are the special features of a dynamic and evergrowing civilization.

The expanding universe of knowledge, material to etherial, does satisfy our intellectual query. Science develops insights into the mysteries of creation and existence which have always attracted the attention of man since the dawn of civilization. Science, besides satisfying the intellectual curiosity of man and providing materials and media for intellectual exercise, has disciplinary effect on the minds of men. We may call it the disciplinary values of science. Since science covers the widest range of knowledge, the learner wonders at the intricacies and mysteries of the universe, the known and the unknown. These tend to create a broader outlook in the mind of the learner. Further, science is universal in character; it has no barrier of any kind. It is international in outlook.

The scientific revolution began in Western Europe where modern science was born but its home is now the whole world. The fruits of scientific discoveries in one country are enjoyed by the people all over the world. Science is not concerned with caste, creed or colour nor recognises territorial barriers. Such a pattern inherent in science will definitely have an impact on the minds of the learners and is expected to help develop broadmindedness in them. The study of science has several other disciplinary values. For instance, science is an interest-awakening subject; its pursuit demands persistent efforts, diligence and patience.

Any experimentation in science requires keen observation, concentration of mind as well as accurate representation of facts. There is no place for prejudice or bias in science. Scientific pursuits warrant objective observation and impartial judgement. Engagement in any scientific activity, be it theoretical or experimental, therefore, presupposes intellectual honesty, perseverance, concentration of mind and broadmindedness. In science we do not conclude or predict anything on the basis of superstition, traditional belief or hearsay, unless the facts are based on proof.

In science there is no place for sentiment or emotion save rationality. A scientific result to be acceptable must be valid for all cases. In pursuing a scientific problem, one has to define the problem, plan the process, collect relevant data, formulate necessary hypothesis, repeat the processes if necessary, apply to specific cases before generalising. During the process, one has to be logical and objective at every step.

Thus, scientific pursuits demand such qualities as minute observation, scientific attitude of mind, persistence, perseverance, concentration of mind, accuracy of measurement, patience; logical, objective and unprejudiced judgement; respect for other's opinion, respect for truth, etc. These disciplinary qualities of mind, if cultivated through the teaching of science, may be carried over to manifest in the general behaviour of the learner. This will prove useful for living as an efficient social individual in the society.

No other subject provides for inculcation of so many disciplinary qualities of the mind of the learners. It is hardly necessary to elaborate the utilitarian or practical values of science. The present world is a world of science and technology. Every thing or every event happening around us demands some knowledge of simple scientific facts or principles. Without the elementary knowledge and information of science, we will be at a loss in the society.

Science is now everyday-science for everybody; its knowledge must no longer be confined to the classes but must reach the masses. The achievements and the benefits of science touch all sectors and all levels of the modern society. The modern man has applied science and technology for the well-being of mankind by inventing machines and by harnessing the resources of nature. The gifts of science have been profitably used for making life comfortable and raising the standard of living. But the use or abuse of the wonderful gifts of science depends on man and his mind.

The recent advances in the field of science and technology and the wide application of the achievements of science in industry, agriculture, medicine, transport and communication as well as their uses in domestic life justify, more than ever, the utilitarian values of science. Science has opened innumerable avenues for pursuing different vocations. A student of science can study engineering and technology, medicine, agriculture or any similar subject and make his career in that profession. In addition, scientific activities have given rise to many varieties of crafts and allied services. Science, therefore, gives opportunities for career-making and pursuing professions and vocations.

In fact, if we refer to "preparation of the individual for the future" as one of the aims of education, then science, as a subject, is rightly serving this purpose. In this age of science and technology there is a demand for technical personnel. The maintenance and creation of new departments, new establishments need the services of engineers, scientists and technicians and there will always be need for research workers in new fields of science.

Educationist Paul Fredman once said, "science is no longer the preserve of a few completely—perhaps abnormally devoted men; it is becoming and increasingly will become, one of the major professions open to any young man of ability, demanding no more in the way of special bent or devotion than medicine or law. But like those other professions, it too will continue to offer a life with characteristic flavour; it will have its own professional standards and its own typical type of thinking and will call forth its practitioners its own loyalties." Science has made a tremendous impact on the cultural life of the present day society which is a product of science.

The thinking, feeling and actions of a modern man are practically guided by the effects of science. There is an involvement of science, direct or indirect, in all works as well as leisure of a modern man. Our habits and attitudes have also been affected by science. The study of science brings behavioural change in the learner and enriches his character and personality.

## Outcomes of Learning Science

Science gives opportunity for creative thinking and constructive imagination. Further, science is a subject where ideas can be experimented upon and verified. The learner develops the habits of searching for the truth. These qualities affect the pattern of behaviour of the learner. The significant aspect of science is that whatever the student learns has immediate application in the world around him. This is educationally very sound. In society, there will always be problems to be solved.

One of the very useful outcomes of learning science is the development of problem-solving skill. If properly cultivated through the teaching of science, the students can apply this skill to solve problems in his personal or social life. One of the aims of modern education is to provide means for utilisation of leisure especially in the industrialised societies. There is no end to interesting pursuits in science, intellectual or otherwise.

Scientific activities provide the best hobbies and pastimes for worthy utilisation of leisure. At higher levels, arts and science are no different. There can be no good piece of art without application of science and on the other hand there is artistic or aesthetic element in all scientific activities.

The great thinkers have always been stressing the unity of science and arts and that they originate from the same root. In the modern civilization, scientific creations glorify arts and aesthetics and science may be said to be the modern substitute for arts in the sense that it is the result of the same kind of creative thought and action which generated arts. Plato clarified this misunderstanding long before the evolution of modern science: "We have certain tendency, on the part of literary and scientific forms, to diverge and develop along different lines, with growing hostility, which suggest fundamental incompatibility . . . this divergence and one-sidedness needs to be overcome.

The origin of both literature and science is the same." Arts and aesthetics are components of culture and civilization. The creation of the universe is a great piece of art. There is aesthetics in the mysteries and harmonies of nature. In the words of an educationist; there is an aesthetic side to the scientist's activities and to his contribution to human culture. On the lowest level he has the satisfaction of adding to the sum of human knowledge; on a higher level he enjoys the subtle pleasure of devising some hypothesis which fits a diversity of facts opening up new areas of knowledge. Appreciation of 'fitness of purpose', the suitability of an apparatus for the job for which it was designed, can give great inward satisfaction.

There is a pleasing skill in avoiding or eliminating sources of errors and in particular the errors of human observation. Wonder is aroused by neatness with which some material quality or some living activity, can be sorted out from other qualities or activities for examination and demonstration. There is an elegance that runs through the logic and handiwork of the scientists. It is seen in the formulae of mathematicians, it is equally seen in the experiments and observations of great naturalists.

The very simplicity of great generalisations of science stirs the imagination. With microscope and telescope the scientist opens up new worlds of wonder and beauty. A speck of living matter becomes a creature of incredible beauty, a snow flake more lovely than diamonds, a distant star becomes a universe. It is at this level that science shares equally with the arts; the privilege of contributing to the aesthetic development of the human race. Culture, in addition to knowledge, includes all activities, thoughts, feelings, attitudes, patterns of individual or social life of men.

## Science and Society

The study of science gives opportunity for the development of favourable traits of human character which become a positive contribution to the cultural life of the society. For instance, with science gaining ground and spreading its influence in the life of man, there has been a profusion of literature based on science.

Science fiction being interesting it adds to the cultural heritage of man. Similarly, the literature on history and development of science is no less interesting. It is the study of the origin and development of civilization itself and has developed into a separate branch of study which contributes to the cultural heritage. The biographies of scientists incorporated in the science course develop a scientific attitude among the learners. The description of the scientists' pursuits, their tenacity and perseverance, are worth reading. Such a study brings out the scientists' attitude towards science and their hopes and frustrations on their way to discovery. Sometimes, even after their invention or discovery, it takes a long time for social acceptance.

The facts about the sacrifices of the scientists for the benefit of mankind stir one's imagination. The lives of Galileo, James Watt, the Curies and others show how the scientists had to suffer to make an original discovery. The lives of the scientists can inspire the minds of the young learners. It is believed that the study of science and the life of the scientists engenders praiseworthy humility.

The study of the scientist's way of discovery is more interesting. It gives the learners an opportunity to grasp the essential steps of scientific method or procedure. For example, the story of the discovery of the Laws of Gravity by Sir Issac Newton or the story of the discovery of the cause of malaria, by Sir Ronald Ross, will help to make the meaning of science clear. It is useful to give the pupils, the idea how scientists sacrifice their personal comfort for the good of society.

Broadmindedness and selfless service to mankind are characteristics of their lives. Science learners can be provided the opportunity for literary expression by being asked to write the details of their observations, procedures and conclusions of scientific experiments they perform. They should be encouraged to write all observations precisely. For example, on completion of a project, an experiment or a field trip or observation of a scientific demonstration, the students may be asked to write an essay describing all aspects of such an undertaking. A science student has to be creative in his writing and will learn the use of right words to describe his experience which is a good exercise in composition and translation of thoughts and feelings into words.

In the study of science, there is scope to do project work and group practical work in the laboratory as well as in the field when a sense of good social habits, attitude of fellow-feeling and cooperation as well as mutual understanding are automatically cultivated. The significance can be well appreciated when the students undertake project activity as a part of studies such as field trips and educational visits. In such activities, the spirit of team work and cooperation is visibly exercised.

A good example of cooperative team work is scientist William Harvey and his group of scientists. In group work, the students learn to work as a member of a group, help each other, think together and plan together. This process of teaching science has important educative and cultural value. In project or laboratory work, the students learn to do manual work and are compelled to shoulder responsibilities. Moreover, the attitude of dignity of labour is cultivated. In practical work of science, each student has to do his own job by way of arranging and setting up experiments and collection of data.

There is no discrimination on the basis of social status or economic condition; rich and poor, irrespective of family background, have to do manual work in the laboratory. A scientist is a seeker of truth and scientific facts give a true picture of nature. In scientific pursuits, it requires intellectual honesty at each step.

In an experiment, one has to record correct data, collect authentic information and make objective interpretation of observations. Anything other than truth will lead to wrong results. For exploration of the unknown, scientists have to proceed carefully on the basis of the true picture at each stage of the process. Intellectual honesty and love for truth coupled with sincerity of purpose are virtues that are prerequisites in any scientific pursuit.

Science as a subject has two very important virtues peculiar to it. The study of science imparts training in 'scientific method' and develops 'scientific attitude' in the learner. These qualities can be cultivated only through this subject which can qualify the learner to live as truly efficient citizen in a scientific society. No doubt, science has established itself as the most worthy teaching subject in the modern world not only for its utilitarian values but also because of its disciplinary and sociocultural values, and above all its empirical and logical approach to the facts.

History records that Galilio, even in the face of condemnation and trial for punishment, never lost faith in his conviction because he was convinced of the truth of his observation. It is said that even though he publicly negated his own scientific statement in the trial chamber, he yet muttered outside "oppur si mouve". Bertrand Russell designates this as the voice of the world. Thus, in science ultimately truth prevails because science is nothing but truth. There can be no better moral value of a subject than this virtue. Hence, science is supreme among all the subjects.

# 3

# Techniques of Teaching Science

Teacher's main job lies to teach. There is not a method or the methods of teaching science which could suit in all the situations. All teachers are different. No two children are alike. Teachers differ in personality. Children differ from family to family and from locality to locality — even in respect of their mental and physical development. Thus, a teaching method is largely governed by these three factors — environment, teacher and pupil. It will be advisable to lay down a series of methods of science teaching, so that the teachers could select from them according to their need.

## Teaching Approach: Selection

Every teaching method is unique. Different teaching methods are not just a matter of fashion or personal whim. There are grounds for choosing a particular method for a particular situation. A teaching method may prove to be suitable for one situation but completely inappropriate for another.

Our teaching method should be compatible to our objectives. Suppose our objectives are:

1. To ensure that the students obtain the factual knowledge of the subject.

2. To give the students an understanding of the basic concepts.
3. To help the students acquire practical laboratory skills.
4. To promote the interest of the students in the subject.
5. To develop the students' ability to communicate his understanding of the subject in their own words.

If we look at these objectives and choose suitable methods to achieve these objectives, we will see that in order to achieve objectives No. 1 and 2 'lecture method' will be more appropriate, but by this method we will not be able to achieve objectives No. 3 and 4, for which perhaps the laboratory method would be more suitable. The lecture method probably will also be suitable to achieve objective No. 5, if during the lecture notes are not dictated, but students take their own notes.

## Trying-out Teaching Skills

You learned several methods for science teaching. You also learned how to select particular method or methods for teaching a particular lesson. Now you are to teach science, and all you learned is mostly theoretical. As a practising teacher, you should practise these methods or atleast some teaching skills before facing the class. One of the most effective, efficient and dependable methods for developing teaching skills is "microteaching." What is microteaching? How can you use this technique for developing your teaching skills? Answers to such questions might help you to make yourself a better practising teacher.

## Microteaching

Practising teachers, when they go to teach in the schools, they face many problems even after completing many theoretical aspects of pedagogy in their course work. Actually teaching is not a simple process. It involves many aspects like class organisation, command on the language, control on the class, probing the students explaining concepts, ideas and so on. The whole idea of microteaching was started at the Stanford University in 1960 by Alien and others. It was initially designed for practising teachers to do practice in teaching in front of their own classmates (it was also recorded on a videotape) before going to the actual classrooms. The feedback given by the peer group and received by the video replay helped the teacher to identify his/her shortcomings and do more practice in identified areas.

The idea of the whole process was to bring improvements in one's teaching. Microteaching was found quite an effective process in developing teaching skills in preservice teachers during their training period. Since then it has gained quite a popularity in various teaching training institutes and other training centres. A large number of researches have been done in the field of microteaching and many microteaching centres and research units have been established. Other than preservice teachers training, microteaching is being used for in-service teachers training and in many professional fields like medicine and nursing, etc.

As the name indicates microteaching means teaching a short (micro) lesson, for a short time (5-10 minutes); teaching to a small group or may be their own peer group or a small group of students, concentrating on one particular aspect of teaching. The rationale for this approach is that teaching is a complex process involving various aspects, microteaching helps in concentrating on one aspect at a time, therefore reduces the situations to a manageable proportion.

The experts and research in microteaching have identified many skills in teaching such as — skill of introducing a lesson, class management and organisation, maintaining students attention, art of questioning, use of chalkboard, lecturing, using the audio-visual aids, reinforcement or encouraging the desired behaviour, explaining things and many more. In microteaching session each skill is practised and mastered individually. Once the various skills have been practised in the simulated situation the teacher will be able to combine all or some skills of teaching in the real situation of classroom teaching.

Before actually starting any practice in microteaching it is important to orient the trainee teachers with the philosophy and technique of microteaching. This could be done by supervisors or other subject experts, then the skills have to be identified for microteaching practice. *One* way of doing this is to plan and give a regular teaching lesson in simulated situation (in front of the peer group).

From this simulated teaching the weak areas of skills can be identified after getting proper feedback, or it can be decided by the subject expert or the supervisor to give practice in some skills which will be more important in teaching, for example, in science teaching, problem solving, enquiry approach and some others can be identified.

The microteaching may be done in the following steps.

### *Microlesson Planning*

After deciding on the skills to be developed the first step is the planning of a microlesson on the identified teaching skill say introducing a lesson on a selected topic for a short period of time say five-six minutes.

### *Practice of the Teaching Skill*

The-second step is the teaching session where the planned lesson on the specific teaching skill will be delivered to a selected number of classmates who can play the role of that level of students for which the lesson has been developed. The rest of the classmates and the supervisor are observers.

### *Feedback Session*

The teaching session will be followed by a short session where the feedback is given by the classmates as well as by the supervisor. At Stanford University the videotape was used for the purpose of feedback which was discussed in the class and by actually seeing himself/herself the teacher was able to improve. In some places video-tape records are being used in microteaching sessions. But it involves heavy expenditure therefore it can be avoided.

*Re-plan Session*

In the light of the feedback received from the students and teachers' the lesson will be replanned.

*Re-teach Session*

The replanned lesson will be retaught to the same group or roles can be exchanged by the students for about the same time.

*Re-feedback Session*

The teacher will again receive the feedback from student-teachers (observers) and the supervisor. If needed the same cycle can be repeated to achieve the desired level of performance. Usually there are two cycles but for some there may be three cycles and for others only one cycle is sufficient.

If some sessions of microteaching are planned and arranged in the training colleges for the science teaching, practising teachers will be better equipped and will go to their practice teaching classes with more confidence.

## Approach to Study

1. You learned several methods of teaching science. You also learned how to select a particular method/methods for a particular topic and situation. You also learned some teaching skills for science class. According to NPE — 1986 "All teachers will teach, and all students will study". Now you will teach science effectively, as you know 'how to teach'. But just teaching is not enough. Your students should also study after you taught them some science. For effective learning, your students should be trained for some study skills, so that they know, how they should study?
2. A student should develop and employ study skills in order to derive the maximum benefit from the teacher's

talk (lecture) along with T-L Interaction (Discussion) in the classroom, observations and experiments in the laboratory, reading books and journals in the library and doing assignments (homework) at home. While some universities abroad have started courses on "Learning to Learn (Development of Study Skills)," atleast we should make a beginning.

3. "How should I study", should be the objective of every student, you or your student. There is a great deal of literature based upon researches on learning, retention, thinking, cognitive styles, reading techniques, information processing, scheme, metacognition, note taking, problem solving, time management and related subjects. We should take its advantage to get the answer of this question. We should discover the difference between teaching-learning of Science in the School of Today and Yesterday.

   In order to complete the Today's Science Course, the class contact hours are less compared to the content to be taught, and hence our students are overloaded. Therefore, Today's Education should provide an opportunity to students, to learn by reflective thinking, to internalise the concepts and to demonstrate creativity. Besides, Today's Education should encourage and motivate students to learn for themselves (self-learning) and pursue lifelong learning.

4. *Subskills for Learning*: There are a number of subskills (ingredients of study skills) for learning. Reflective Thinking constitutes the core. Listening (Paying Attention) is "indeed the starting point of studying in the classroom. Likewise, Reading Skills are basic to self-study. Note Taking, Question-Answer Technique, and Communication Skills greatly add to the learning process. Knowledge about Information Processing and Schematisation may enable - to adopt more effective study skills.

5. *The Learning Process:* How students learn, retain and recall information has been the subject of intensive research by

cognitive psychologists. The more active the learning is, more easily connections between new information and already existing information are made. More connections facilitate easy recall and longer retention. Meaningful conclusions can be drawn from the findings of the cognitive research studies. Some of them are as follow:

- Readiness to learn and motivation to learn are most essential pre-requisites to all Learning in order to generate your own motivation.
- Objectives of learning, i.e., the intended capabilities after learning must be kept in view. It saves great deal of time and effort to go by objectives.
- Variation of activity during learning engages you in learning for a longer time.
- Incremental, i.e., step-by-step learning or learning in small chunks, and self-evaluation during each chunk of learning ensure more permanent learning. Each chunk should have one new idea.
- Thinking while learning ensures your involvement in the process of learning. Mere note taking is not sufficient. Try to understand what you are learning. Recall previous topic. Look for links (connections) between present topic and the previous topic. This reinforces learning on both the topics.
- Memorisation and recall of facts and figures is often aided by remembering cue words, flow diagrams, acronyms, etc., e.g., acronym VIBGYOR aids in recalling the seven colours of a rainbow.

Sustained research on learning shows that you cannot continue to learn at a stretch. Nor is the rate of learning constant with time. In a classroom, the period of time, for an adolescent, over which the learning is sustained with a reasonable rate of learning, i.e., attention span of learner is about 20 minutes. So after every 20 minutes the teacher should change the activity for effective learning. They may use different teaching aids, tell a joke, make a mistake, ask a question, give a quiz test, etc.

6. *Memorising Facts and Figures*: Whatever innovations in teaching and learning may take place, one thing is sure you will have to memorise atleast some facts and figures. You cannot afford to device equations and think but all possible facts in the examination hall. There are a number of learning strategies for memorising. Some of them are as follows:

   - Reviewing, repeating or rehearsing the facts are necessary. Repetition and restatement in your own words help in conceptualising and remembering things.
   - Never attempt to memorise what you do not understand. For your understanding explain something difficult to someone else.
   - Clustering, categorising and grouping of facts are excellent aids to memory.
   - Making acronyms, one words, and analogies may aid recall. Different students prefer different techniques of memorising. You must also have memorised many facts and figures, and so your friends. Share your experiences. What techniques you have used for memorising. Discuss. You can discover your own preferred methods for memorising.

7. *Questions during Lectures*: Lecture Method is not the best method of teaching Science. What, then can you do about it if your teacher decides to use Lecture Method? The least you can do is to ensure you get the message during a lecture. When in doubt, always raise your hand and pose question. If your teacher asks questions during lecturing, raise your hand to show willingness to respond. Do not miss this chance. And when you as a teacher lecture, encourage your students to ask questions. You also ask questions to ensure that your students are

understanding what you are teaching. Have a healthy Teacher-Learner Interaction (Discussion) during your lecture.

8. *Reading Skills*: As a science student you are to read a lot. If you develop some Reading Skills in you and then in your students it will help a lot in learning science.

   A reading strategy called ***SQ3R*** technique is frequently advocated for effective reading.

   ***S*** Surveying the subject matter

   ***Q*** Questioning oneself about the need of reading

   ***R*** Reading and deriving meaning

   ***R*** Recalling the previous learning

   ***R*** Reviewing the salient points

   Another method of effective reading is ***MURDER*** Technique.

   | | |
   |---|---|
   | ***M*** Mood | Setting the mood for study doing something relevant and then switching over the study. |
   | ***U*** Understanding | Making important and difficult ideas. |
   | ***R*** Recalling | Material without referring to the text. |
   | ***D*** Digesting | Internalising the material. |
   | ***E*** Expanding | Knowledge by self-questioning. |
   | ***R*** Reviewing | Learning from mistakes. |

9. *Note Taking*: There are many good reasons for taking notes while listening to a teacher or reading a book. Note taking promotes listening, thinking, deriving meaning, concentrating, and helps in keeping record for future reference.

   Different students adopt different techniques of taking notes. Some students write in the margins of the books, some on cards and others on proper notebooks. Some students write cue words, some half sentences, others make scheme maps, i.e., cue diagrams beside key words. One needs not write what the teacher speaks; one may also not write what the teacher writes on the chalkboard or what he shows through overhead projection. One may develop notes in one's own style of adding-on-information in blocks, flow diagrams, schematics, etc. After all, one's notes are primarily for oneself. Good note taking is, as much a process of learning as good notes are a product of learning.

It is recommended that students should use note sheets with adequate margins for future additions, recognition, revision, arrows, etc.

A recent malpractice is to photocopy someone else's class notes. It does not work. One's own notes are the best. In your own notes, you can build links to your previous knowledge, you can create your own memory aids — these are more powerful for you than other people's aids, your own notes can concentrate on those things which you find difficult to understand.

10. *Communication Skills*: You have a great deal of scope for developing oral and written communication skills in your professional studies. How are these skills acquired?

    - *Skills of Oral Communication:* One should learn how to ask a question (Question Skill), how to be precise, how to respond, how to offer a suggestion, how to propose a solution, and how to speak in public. It is necessary to listen when you speak. One should contribute in group discussions through critical thinking and quick responses. Quite often, someone else has just said what you wanted to say. Why not you? At times, take a lead. Tutorial classes are friendly meetings with free flow of ideas. You should take advantage of your tutorials to develop skills of oral communication in you.
    - *Skills of Written Communication:* Assignments and laboratory experiments should be written up

independently. While writing, think visuals; draw graphs and other diagrams to convey your thought effectively. Writing is the process of expression of thoughts in black and white. The term black and white includes written words, figures, tables and the use of desk computers. It is necessary to develop writing skills because you need to write out notes, project proposals and reports, executive summaries, critical appraisals, etc., all your professional life. Remember: 'A picture is worth more than thousand words.' So try to put words in a visual form by making effective diagrams.

11. *Problem Solving Skills:* One of the best ways to learn how to solve a problem is to ask someone else to explain how he proceeded to solve it and why. A heuristic approach to problem solving suggests such strategies as:
    - Thinking aloud
    - Working backward
    - Using trial and error
    - Considering other possibilities
    - Using means for analysis
    - Subgoaling and working by parts, etc.

    Practice in solving problems leads one to think creatively, to employ different techniques and to arrive at acceptable solutions. Consider the formal type of problem solving that we do in science, e.g., problems like 'Prove that. . ./ 'Show that. . .,' etc., in which case the answer is known. In problems like 'Evaluate. . .,' 'Determine. . .,' etc., you have to check your solution by answering some questions to yourself, like:
    - Is the answer sensible?
    - Does the answer fit in the original data?
    - Is the answer unique? Would any other answer fits just as well? If so, how can I decide which is more likely to be correct?

12. *Time Management*: You are to do co-curricular activities along with curricular activities — assignments, homework, self study. You have to do all this plus a lot more of your own. You have just 24 hours in a day. You cannot increase that. So you are to manage your time. There is no cut and dried formula for time management. You are to lay down your own priorities for the work you are to do and complete.

    It is worthwhile to be regular in homework, and complete your assignments and projects well in tune to avoid undue tension relax. A relaxed mind learns much better. Time for relaxation should be budgeted in the work plan. Make your daily, weekly, monthly and annual work plans, and then stick to the schedules of your plans. This will make your learning a joyful learning.

13. *Learning Environment*: It is for you to create an environment conducive to your learning. The best environment is that where you as an individual feel motivated and reasonably comfortable to study. Reasonable comfort includes your body posture with no or minimum stress, lighted study corner with fresh air and controlled dust, heat and noise. Concept of noise may differ. Some students can study perfectly comfortably with radio or music deck at full volume, some like it in the background and others cannot bear it at all.

14. *Preparing for Tests and Test Taking*: Having developed study skills and even after reading and preparing for a test, one may not still score well in a test. This is because Test Taking is a reverse process compared to the acquisition of knowledge. In a test, one has to recall, recognise, manipulate data, compute, analyse and write. Not only this, one has to adopt a strategy of scoring.

    (a) *Pre-examination Preparation:* For pre-examination preparation, complete any outstanding work as quickly as possible, even if this means sacrificing the quality somewhat. Your notes are the most helpful tools for preparation. Organise your notes and highlight the key points. Look at those problems particularly which you do not know well. Answer the questions of the past years question papers. Explaining a concept to someone else makes you understand better. All this will give you a good deal of confidence.

(b) *Test Taking:* You should know the art of Test Taking to score more. ***SCORER*** will help you to score more.

| | |
|---|---|
| ***S*** stands for | Scheduling the time |
| ***C*** stands for | Cue words |
| ***O*** stands for | Ordering of questions to be attempted |
| ***R*** stands for | Reading between the lines |
| ***E*** stands for | Estimation of answers |
| ***R*** stands for | Review and Revision. |

16. *Relax*: Relaxation is Basic to good study skills. One can think better and respond better when relaxed. Once you are able to think/ may it be in classroom, laboratory, field or at home, you will be able to grasp the subject. Some knowledge about the process of learning, information processing, note taking, etc., discussed here would lead you to better organisation and management of learning.

## Task Method

"Believe nothing because you have been told about it. Don't believe what your teacher tells you merely out of respect for the teacher."

Who said this? When this question is put to science teachers in seminars the reply is 'some scientist'. Actually it is the statement of Lord Buddha.

Many times teachers teach and children do not understand. This is because children are conditioned in accepting and memorising what teacher tells them, without questioning and understanding.

In science there are many facts, principles, laws, concepts and skills, which children memorise without understanding. Have you ever thought why is it so?

***Cognitive Development of Children:*** There are three types of children:

1. *Formal Operational:* They can understand science without activities.
2. *Concrete Operational:* They can understand science only if they do activities by their own hands.
3. *Pre-operational:* They can understand science if they do activities repeatedly by their own hands.

This is the way Piaget categorised the children. These are the children's cognitive stages of development.

***Example***: Siddiqis worked with Piaget at Florida State University (USA) in 1970 when he came as a Visiting Professor. He went to the neighbouring schools to demonstrate how children at these cognitive stages look at things, think and response. There it was found that only 5 per cent children at primary level (I-V) were at formal operational stage, the remaining 95 per cent children were either concrete operational or pre-operational.

By using Piagetian tasks, Siddiqis conducted Research Studies on 1206 Primary school children in Delhi from all type of schools during 1975-77. The average interview time with each child was about 15 minutes. So it took about 300 hours to complete the task. It was found that only 4.4 per cent children at primary level (I-V) were at formal operational stage. In Class V formal operational were 9.4 per cent, concrete operational 69.3 per cent and pre-operational 21.3 per cent.

Even at Upper Primary Level (VI-VIII) a good number of children will be at concrete operational stage. This shows that specially at primary and upper primary levels, if children are not involved in doing activities or get first-hand observations, the concepts do not become clear. They may memorise and recall without understanding.

When the concepts are not made clear in lower classes, their cumulative effect is seen in higher classes. Other than this there are new concepts at secondary and senior secondary levels which can be clarified through activities or relevant demonstrations.

**Working Science**

Generally in our schools, science is taught by giving Lectures, and sometime by demonstrating few experiments. Children are not really involved in doing activities by their own hands. Teachers should also learn a lesson from the following saying:

"I heard and I forgot,
I saw and I remembered,
I did and I understood "

This is not just a saying. As hypotheses these were tested by Siddiqis and they were found correct.

If our objective is that the child may remember the content for sometime. Lecture Method may be relevant without even using Audio-Visual Aids. If the objective is that the child may remember for longer duration. Lecture will not work. But Lecture-Demonstration will be more effective. Care should be taken that every child is able to observe what is being demonstrated using all the 5 senses (see, touch, smell, taste, hear).

But if the objective is that the child must understand then doing is the only method for pre-operational and concrete operational children. Lecture and Lecture-Demonstration are not going to work. The children will have to be involved in doing activities by their own hands.

Thus, Activity Based Science-Teaching is the only way of teaching science, if we want our children to learn science. Learning science is understanding science, and not just memorising.

Sample science Activities for primary, upper primary and secondary classes have been given. Some of the activities from this list may be selected according to the science topics to be taught by you in various classes. These activities may be demonstrated by you, or may be done by your students by their own hands, in science classes. These activities will make science teaching quite interesting and learning science by Activity Approach will be a fun for students.

## Process Mode

Another important aspect of Activity Based Science Teaching is Teaching with Process Approach.

"If you will teach content, then who will teach science" D.S. Kothari. Science is not just content. Science is content plus something. That something is Process. If our stress shifts from content to process, content will automatically be included and much more meaningfully.

American Association for the Advancement of Science (AAAS) has identified 13 science processes — observing, classifying, using numbers, measuring, using space-time relationships, communicating, predicting, inferring, defining operationally, formulating hypotheses, interpreting data, controlling variables and experimenting.

If we teach science through process approach, it automatically becomes activity based. If a child learns science through process approach, he learns lot of science on his own. Thus, Process Approach is one of the best ways of teaching science, as it motivates the children, and they get involved in doing activities.

Certain competencies are to be developed in children while teaching science like spirit of enquiry; objectivity; courage to question; problem solving; decision-making; investigating; developing scientific attitude or temper; using scientific method; reducing all sorts of prejudices based on sex, caste, religion and language, etc. " These are also the objectives of NPE — 1986. These competencies may be developed in children through Process Approach of Teaching Science, more meaningfully.

Although one has to devote a little more time when teaching science through Process Approach, but teaching-learning process becomes more interesting and effective. There are teachers who use Process Approach for teaching science in their classes, and for their children learning science is fun.

## Environmental Mode

Use materials (like plants, animals, human body, living-non-living, natural-manmade, wind, hails, snow, dew, etc.) and phenomena (like rains, lightning, day and night, seasons, dispersion, rainbow, photosynthesis, *etc.)* of the environment around you (like family, house, neighbourhood, school, earth, sky, human body, etc.) as teaching aids, when teaching science. Involve students in discussions based on what they observe in their environment in and outside classroom.

Make an exhaustive list of materials and phenomena available in your environment. Add more items whenever you see some more, in your list. Classify them according to the topics you teach in science, and use them again when teaching science. Your students will enjoy learning and you will enjoy teaching science by EVS approach.

## Utilisation of Various Methods

About 1000 Science teachers from 250 Schools of the country (12 states) were oriented in Teaching of Science in the 'In-service Teachers Training' workshops during (1997-2000) conducted by D.M.N. Siddiqi.

***Himachal Pradesh:***

1. Kangra 2. Mandi 3. Chandigarh

***Punjab:***

4. Bhatinda 5. Jallandhar

***Haryana:***

6. Hissar 7. Ambala

***Bihar:***

8. Ranchi 9. Patna

***Uttar Pradesh:***

10. Brij Vihar

***Rajasthan, Gujarat:***

11. Padma Biriani

***Andhra Pradesh, Tamil Nadu, Karnataka:***

Hyderabad

***Delhi:***

13. Delhi SCERT, Delhi DIETs, Municipal Corporation Delhi (MCD) Science Centres and Research & Training Institute.

***All the 22 Methods were tried out:*** The following 11 methods were found most effective. When teachers taught, children also learned (understood).

***Child Centred Approach (CCA)*** — for teaching Science Concepts.

***Teacher Centred Approach (TCA)*** — for teaching Science Skills or Processes.

***Scientific Method*** — Problem, Hypotheses, Experiment, Conclusion.

***Process Approach*** — observing, classifying, using numbers, measuring, using space-time relationship, communicating, predicting, inferring, defining operationally', formulating hypotheses, Interpreting data. Controlling variables and experimenting.

***EVS Approach*** — Our Body, Our family, Our House, Our Neighbourhood, Our School, Our Earth, Our Sky.

***Discussion or Interactive Approach***

(i) Teacher — Child Interaction (TCI)

(ii) Child — Child Interaction (CCI)

(iii) Teacher — Material (Teaching Aid) Interaction (TMI)

(iv) Child — Material Interaction (CMI)

***(Improved) Lecture*** – add 6 [(i), (ii)]

***(Improved) Lecture*** – Demonstration — add 6 [(i), (ii), (iii), (iv)]

***Self-Learning Methods:***

- Self-Learning Modules
- Programmed Learning Technique, (now is becoming old, but still effective)
- CDs (Computer Discs)
- Multimedia Packages (Printed Material, Audio-Tapes, Videotapes, CDs, Internet, etc.).

***Concrete to Abstract Approach***. Direct purposeful experiences. Contrived experiences. Dramatic participation, Demonstrations, Field trips or Excursions, Exhibits, Motion pictures and Television, Radio recordings and still pictures. Visual symbols and Verbal symbols.

***Activity Based Teaching Approach***

***What do we mean by Understanding?*** This Question was asked by a good number of children learning Science at various levels. Their answers are summarised below:

I feel, I understand Science because:

1. state it in my own words.
2. give its examples.
3. recognise it in various guises and circumstances.
4. see connections between it and other facts or ideas.
5. make use of it in various ways.

6. foresee some of its consequences.
7. state its opposite or concourse.

***Strengths, Weakness and Suggestions for Further Improvement.*** The participating science teachers gave their feedback for the teachers training programmes. This is summarised below:

- We learnt several new techniques for teaching of science which we were not taught in our B.Ed. Programme. Now we are better equipped for Activity Based Science Teaching. When we will use the methods we have learnt here in our science classes, our students will hopefully understand better and enjoy science more.
- (Improved ) Lecture and (Improved) Lecture - Demonstration seem to be more effective than just Lecture and Lecture-Demonstration methods.
- Now we know that for teaching science Concepts we should use LCA, and for teaching science skills/processes we should use TCA.
- We think that the LLI, the EVS Approach and Scientific Method are the best techniques for science teaching.

***Concluding Remarks.*** Our Science courses are mostly content-oriented. Whenever the courses are updated more content is added. Courses become heavier and heavier while time allotted in the school timetable remains just about the same. Thus, students have to cover much more content in comparatively much lesser time. Moreover classes are crowded and teachers are heavily loaded. Thus, in spite of sufficiently available Science equipment for experiments and demonstrations, the content is covered by asking students to read books or giving lectures to cover more content in lesser time without demonstrations and experiments.

Under such classroom conditions we have to devise and use some self-Learning techniques in teaching Science. These along with some other researched methods/approaches LIKE

Learner Centred Approach, Teacher Centred Approach, (Improved) Lecture Method, (Improved) Lecture-Demonstration Method, Concrete to Abstract Approach, and Activity Based Teaching Approach will be more effective in science teaching.

## Scientific Mode

Scientific method is one of the methods for discovery. Training in scientific method necessitates discouraging traditional demonstrations and laboratory experiments. When teaching by scientific method, the students do not perform the prescribed list of experiments in which emphasis is on the verification of some already known science principles and laws. Instead they are involved in investigatory problems whose solutions are not available either in textbook or lab manuals.

***Steps of Scientific Method:***

1. *Problem:* What does a student want to learn?
2. *Hypothesis:* What does a student think the answer would be?
3. *Experiment:*
   - Testing of Hypothesis by an Experiment. How could a student tell the answer would be what he predicted?
   - Collection of Data
   - Interpretation of Data. What happened when a student tested his hypothesis?
4. *Conclusion:* What do the results of the experiment show? Is the hypothesis accepted, rejected or needs modification.

***Advantages:***

1. Students do their own learning under the guidance of their teacher.
2. They learn to propose, formulate and structure problems.
3. They learn to collect varied pieces of information relevant to the problem from different sources.

4. They collect the evidence to prove or disprove the identified hypotheses.
5. They learn to solve everyday problems.
6. They are more closely familiar with various things, their applications and relationships instead of having mere knowledge.
7. They establish a healthy and favourable relationship with their teachers.

***Disadvantages:***

1. This method is very slow, long and time consuming. Therefore, the syllabus (knowledge and information) could not be covered in the allotted time. Thus, a different syllabus is needed for this method as in some American science programmes like PSSC, BSCS and CHEM Study of which formulation of hypotheses and their testing is the dominant feature.
2. There is too much emphasis on practical work which may give a wrong concept of the nature and philosophy of science in general. Learning science is a joyful process but too much practical work may make it dull and routine type of affair.
3. Most of the teachers are perhaps not able to teach by this method, as they have not practised teaching science by this method.
4. All students are not capable to learn by this method.

## Mode of Project

A project is any purposeful activity. It may or may not involve investigation. If it involves investigation it will be investigatory project. In progressive schools students are allotted some projects during summer holidays. It may be a curricular or enrichment activity. Teacher try to cover a part of syllabus by this method. Parents also help their wards when they work on their projects. Titles of some projects generally given to students are as follows:

1. Indian Scientists;
2. Story of Wheel;
3. From Earth to Moon.

Here students read some reference material, talk to some professional people and the public, observe what is going on around them and so on. This is how they gather data. Then they analyse their data, and prepare their project reports. The report may be hand-written or typed and it may include pictures, charts, graphs, models and so on. Students learn a lot when these project reports are discussed in the Class. Project Method is an example of Assignment Method.

## Doctrine for Choosing Teaching Techniques

There are some guiding principles for determining our teaching methods. They are as follows:

1. Every method we use should be based on an understanding of the pupils in our classroom, and not on an understanding of children we have read about.
2. Every method we use should be one which we thoroughly understand and believe will be successful with the pupils in our classroom and not merely one which we have heard is successful.
3. Every method we use should lead the pupils to a sense of achievement through interest and purpose.
4. Every method we use should stimulate the pupils to think and cooperate actively.
5. Every method we use should be based on the realisation that the word education derives from a Latin word 'educo' which means "I bring up" or "I nourish." Education is a "drawing out" and not a "putting in."

Mindful of the above points there are certain vital instincts which stimulate the child to learn and through which he acquires knowledge and skills. They are instincts of play, imitation, curiosity and competition. These are the instincts which are principally concerned with the educational development of the child and which we must consider when planning our methods of teaching science.

Beside these vital instincts from which the impulse to learn springs, there are some physical agents which enable a child to acquire skills and knowledge. They are as follows:

1. The hands — Learning by touching and doing.
2. The eyes — Learning by seeing.
3. The ears — Learning by hearing.
4. The mouth — Learning by saying.
5. The nose — Learning by smelling.
6. The tongue — Learning by taste.

These senses should be utilised as much as possible in our method of science teaching for creating effective teaching learning situations

## Several Techniques

### *Traditional Methods:*

1. Lecture Method (8 4)
2. (Improved) Lecture Method (8 5)
3. Demonstration Method (8 6)
4. Concrete to Abstract Approach (8 7)
5. Lecture Demonstration Method (8 8)
6. (Improved) Lecture Demonstration Method (8 9)
7. Laboratory Method (8 10)

### *Discovery Methods:*

8. Scientific Method (8 11)
9. Project Method (8 12)

10. Investigatory Project Method (8 13)
11. Problem Solving Method (8 14)
12. Heuristic Method (8 15)
13. Inquiry Approach (8 16)

***Other Methods:***

14. Discussion or Interactive Approach (8 17)
15. Programmed Instruction Method (8 18)
16. Multimedia Package Approach (8 19)
17. Individualised Instruction Method (8 20)
18. Modular Approach (8 21)
19. Teacher Centred versus Child Centred Approach (8 22)
20. Activity Approach (8 23)
21. Process Approach (8 24)
22. Environmental Studies (EVS) Approach (8 25)

## Lecture Method

This is one of the most popular methods of teaching in our schools. This is a teacher structured method, and the students are just passive listeners most of the time. Very few teachers allow questions during the lecture, though some of them give some time to their students to ask questions after the lecture. Many students forget their questions by the time the lecture is over. So their questions remain unanswered. Teachers talk most of the time without using any teaching aid though some of them use blackboard.

For them the lecture method is talk and chalk method. Sometime the back benchers cannot even read what is written on the blackboard. Students take notes of the lecture, and slow writers miss many points.

Research studies have been made on what actually happens in classrooms where teacher is teaching by lecture method. 'Step into a classroom and what do you hear. The chances are better than 60 per cent you will hear someone talking. If someone is talking the chances are that it will be the teacher more than 70 per cent of the time. The teachers talk more than all the students combined.

The teacher is the main actor 85 per cent of the time. Pupils receive recognition, praise and encouragement only 1-2 per cent of the time." According to Flanders, N.A. "We want to encourage our pupils to learn to express themselves fluently, yet on average there are two speeches by each pupil each day in class, and the mean length of these is 8.4 words."

Students dislike lectures. In an adult education class those who had heard only 15 minutes of a radio talk could score 41 per cent on a factual recall test while those who listened to the 30 minutes of the lecture could score only 25 per cent on the same test. In other words, the second period of 15 minutes proved to be for less productive as it reduced the listener's ability to recall the first 15 minutes. It may be due to fatigue set in which students lost the concentration. The span of concentration seems to be less for the child than for the adult. Children switch their attention rapidly from one subject to another.

It implies that the teacher should not lecture continuously for more than a very limited time, say 15 minutes with an average secondary school class. After that a change of activity, to practical work, watching a film, writing or working in small groups, or discussion could be submitted.

Lecture method has some advantages and disadvantages.

***Advantages:***

1. It is quick, and a lot of knowledge can be imparted in quite a lesser time.
2. It is highly efficient if teacher teaches in systematic and logical manner.
3. It is convenient and easy and a teacher is free to develop his own style of teaching.

***Disadvantages:***

1. Students' involvement and participation is nil or quite less.
2. Students' previous knowledge or entry behaviours are usually not taken into consideration.
3. Developing scientific skills are neglected.
4. Teacher needs a lot prior knowledge to prepare a lecture.

It is wise to have notes (in points) available while lecturing for ensuring correct sequencing of ideas and figures, but word by word reading from notes should be avoided. Listening to your tape recorded lecture or comments of your colleagues on your lecture will enable you to find out the limitations of the lecture method as well as to enable you to improve your further lectures.

## Lecture Mode in Advanced State

Lecture Method is perhaps as old as appointment of the first lecturer in our country, as lecturer is one who lectures, and uses Lecture Method in his class. But old is Gold. It is correct that in Lecture Method, no audio-visual aids are used, and we do not need audio-visual aids all the time. If our students have enough experience of what is to be talked about in we do not need to use audio-visual aids in the lesson, and the lecture can be improved (enriched) by adding Teacher-Learner Interaction (TLI) and Learner-Learner Interaction (LLI), into it. Involve your students into your Lecture with the two interactions — TLI and LLI. Adding these two interactions with the lecture, some of the disadvantage of the lecture (8.4) can be removed, and the lecture can be improved.

Encourage your students to ask questions (NPE 1986). You should also ask questions to all types of students (average, below average and slow learners) and not just the above average during your (improved) lecture. Sometimes ask some open question, for which first students had to have LLI, before answering.

## Presentation Mode

Demonstration means 'to show'. In lecture method, the teacher just talks but in demonstration method he also shows and illustrates certain phenomenon and applications of abstract principles through demonstration of experiments. Demonstrations provide concrete experiences to students. Demonstrations may also include use of films, slides, overhead projector or the micro-projector.

While using the demonstration method the teacher can:

1. provide significant, rich and worthwhile experiences to students enhancing their power of observation;
2. illustrate abstract ideas in concrete form;
3. coordinate laboratory work with theory and highlight safety methods;
4. pose a problem to students and gather their hypotheses based on their experiences;
5. provide concrete experiences for the solution of a problem;
6. add additional demonstrations after its review, if considered necessary;
7. do those demonstrations which are dangerous for students to perform themselves.

***Characteristics of a Good Demonstration:***

1. It should be clearly visible to all students, even to the back benchers.
2. It should be pre-tested so that it may be convincing, striking and clear-cut.
3. Children should also be involved.
4. Results should not be known to students.
5. If there are more than one demonstration in a lesson they should be well-spread.
6. In a demonstration only one idea should be taken at a time. Too many ideas in one demonstration may confuse the children specially the young ones.

***Points to be Noted***

1. For large scale demonstration, the teacher must spend a good deal of time in thinking and devising experiments to make principles quite clear, and in showing various types of experiments necessary to clarify the numerous difficult concepts of the science course.
2. The teacher should carry out the experiment in such a way that the students should learn how to carry it out by themselves.
3. When carrying out demonstration experiments the teacher should make sure that everything is in order before the lesson begins. He should be certain that the demonstrations will prove his points. The table should be well illuminated either by day light or by electric lights. The background should preferably be white or black so that the apparatus and other materials are clearly visible.
4. The teacher should see the general order and tidiness of the demonstration table. Nothing looks worst than a demonstration table litered with books and other material not to be used. The demonstration table should have only apparatus and materials relevant to the lesson.
5. In this method students are placed in a passive role which may be frustrating or boring for them unless the teacher involves them in thinking behind the experiment and in the manipulation.

This problem can be minimised if the demonstration is assembled ready for action before the lesson and there is every expectation that it will work. Demonstrations need to be rehearsed. For effective demonstration the following questions should be kept in mind:

1. Is the apparatus clearly laid out from where the students view?
2. Do instruments have large enough dials to be read by all the students in the class?

Overhead projector has made it possible to modify some demonstrations so that they are more easily seen and students can take readings and plot graphs while demonstration is proceeding. Under such conditions the teacher can involve their students in making predictions, speculations about causes, and thinking of ways of testing ideas.

*Advantages:*

1. There is a possibility of using more sophisticated apparatus.
2. More difficult experiments can be undertaken.
3. More hazardous experiments may be attempted.
4. Expenses may be minimised compared to laboratory method.
5. There is a possibility to demonstrate manipulative and allied practical skills.
6. There is possibility to draw the attention of all the pupils of the class simultaneously.
7. It takes less time compared to laboratory and other innovative methods.
8. This method is more efficient in a way compared to laboratory method as a teacher is more competent than students to handle apparatus.
9. All students can see the same operation and techniques simultaneously.
10. Teacher is in a position to explain each and every step and to ensure that all students see and interpret all the work in the same manner.

*Disadvantages:*

1. All students do not experiment with their own hands. It is a substitute for laboratory work.
2. When the demonstration is complex or there are too many demonstrations in one lesson, students feel difficulty in understanding the basic concepts, principles and skills.

3. Various details of the apparatus, significant reactions and other essential steps undertaken by the teacher in drawing conclusions are not necessarily visible to all the students of the class equally well.
4. It deprives students of many of the advantages of laboratory method such as handling of the apparatus and other materials as well as making their own interpretations.

## Indefinite Method

Whenever you plan a lesson, ask yourself whether you really need audio-visual aids, as sometimes you do not need audio-visual aids. If your answer is "yes" then ask yourself of what type of audio-visual aids. Edger Dale has arranged the various audio-visual aids, from more concrete in nature to comparatively more abstract (5.4)

1. Direct Purposeful Experiences (5.5)
2. Contrived Experiences (5.6)
3. Dramatic Participation (5.7)
4. Demonstrations (5.8)
5. Field Trips or Excursions (5.9)
6. Exhibits (5.10)
7. Motion Pictures and Television (5.11)
8. Radio Recording and Still Pictures (5.12)
9. Visual Symbols (5.13)
10. Verbal Symbols (5.14)

This will help you to select right type of audio-visual aids for your lessons. This arrangement should not be considered as rigid divisions or watertight compartments.

## Lecture Presentation Mode

This is a method generally used by our science teachers. Students cannot learn science either by lecture or by demonstration used in isolation. Even most effective demonstration cannot guarantee learning in science. This may also be true for the lecture-demonstration method. It can succeed when lecture is based upon concrete experiences of the students' environment and a demonstration focuses them into scientific phenomenon, while discussion between teacher and students goes on in a permissive atmosphere. Demonstration method when combined with a well directed discussion is a successful teaching technique.

It can be modified by allowing limited students participation and by problem solving at times. It fits well into the regular method of uniform class procedure for all students but does not permit of very wide individualisation. If widely individualised and modified, it approaches to a method usually called the problem solving or project method.

Through contrived situations in this method students themselves form concepts. Therefore, lecture-demonstration method should achieve the advantages of both lecture and demonstration methods, and minimise their disadvantages. Lecture-demonstration method though widely used in our science classes has a limited role in the teaching-learning process. This is because all students do not participate actively.

***Advantages***: Lecture-demonstration method, has all the advantages of a demonstration method plus one more that the teacher can also involve the students in discussion.

***Disadvantages***: Disadvantages of a lecture-demonstration method are the same as of demonstration method.

## Lecture Presentation Approach in Advanced State

Like lecture, Lecture-Demonstration Method (8.8) can also be improved. Here along with Teacher-Learner Interaction (TLI) and Learner-Learner Interaction (LLI) two more interactions, Teacher Material Interaction (TMI) and Learner-Material Interaction (LMI) are to be added in Lecture-Demonstration (8 8). Adding all the four interactions — TLI,

LIL, TML and LMI with the Lecture Demonstration (8.8) some of the disadvantages of the Lecture Demonstration (8.8) can be removed. TLI and LLI will improve the Lecture part, and TMI and LMI will improve the Demonstration part of the Lecture Demonstration. This will be (Improved) Lecture Demonstration Method, much better than the Lecture-Demonstration Method of Teaching.

When demonstrating involve students with you instead just teacher demonstrating (TMI) involve also students individually or in small groups in handling the equipment and doing the activities themselves (LMI). During TMI and LMI involve students in TLI and LLI just like you do in (Improved) Lecture. Thus, using (Improved) Lecture-Demonstration method students will understand and enjoy the lesson much better.

## Laboratory Methodology

In this method students perform laboratory experiments or laboratory exercises by their own hands individually or in small groups. So here they are more active and involved unlike demonstration method, where teacher was performing experiments involving few students and most of the students in the class were just passive observers.

Unless students carry out experiments by themselves, they will never get to know what science really is. In this method they get an opportunity to do experiments individually or in small groups. In our schools set up there is a provision of separate science labs but in the West like in USA in most of the schools there are classroom-cum labs.

There is a provision for 55 to 60 minutes class periods. It is upto the teachers, they can utilise this time for theory or practical or both. They carry the needed equipment and material in the classroom-cum laboratory in trolleys, students pick up the material from trolleys, take it to their work tables, and return it to trolleys after doing the laboratory work.

In our schools at secondary and senior secondary stage students go to science labs to do practical work. Usually in every school there are three science labs — Physics, Chemistry and Biology. For lab work two continuous periods (of about 30 minutes each) are given. At secondary stage students go to lab work once a week. If one week they do Physics, the second week they will do Chemistry and the third week they will do Biology, and the same cycle is repeated.

Thus, for each subject (Physics, Chemistry and Biology) the turn comes every three weeks. At senior secondary stage students go to lab work twice a week in every subject — Physics, Chemistry and Biology. Usually students work in groups of about 20. If there are 40 students in a class, they are divided into 2 groups (if one group is doing Physics, the other will be doing Chemistry or Biology) so that the teacher may supervise their students work (in a smaller group) more satisfactorily as well as to guide them individually or in still smaller groups.

Generally lab experiments are of five kinds:

1. Experiments to illustrate scientific principles
2. Experiments to find numerical results
3. Experiments to produce something as preparation of gases or biological slides.
4. Experiments to verify experiments carried out by other scientists.
5. Original work or investigatory science projects or open-ended experiments.

In our science courses there are usually the first four types of experiments. Now a little importance is also being given to the fifth type, the investigatory experiments, at senior secondary stage. Usually there is no coordination between theory and practical work.

The science courses in USA like PSSC, BSCS and CHEM study have employed a more heuristic approach with the practical work embedded right in the middle of the theory. A cycle of events is envisaged.

The pupil is given a laboratory problem and his observations give rise to discussion and hypotheses. These hypotheses are tested by experiment, and so the cycle ends where it started at the laboratory table.

Open-ended experiments, are more difficult to organise and supervise than the traditional ones. The heuristic approach (8.15) demands that the pupils should be allowed to stop and think, discuss and suggest modifications for further experiments.

***Objectives***: When teaching science by laboratory method, the following objectives should be kept in mind:

1. To develop manipulative skills.
2. To arouse and maintain interest in science.
3. To encourage accurate observations and careful recording.
4. To make biological, chemical and physical phenomenon more real through actual experience.
5. To be integral part of the process of finding facts by investigation and arriving at principles.
6. To train the students in science processes and scientific method.
7. To give training in problem solving. To verify facts and principles already taught in theory.
8. To elucidate the theoretical work so as to aid comprehension.
9. To prepare the students for practical examination.

***Advantages:***

1. Learning by doing.
2. Opportunity to handle materials.
3. Learning to follow directions carefully.

4. Learning skills in performing experiments, recording observations and results, summarising data and drawing conclusions.
5. Opportunity for critical thinking.

***Disadvantages:***

1. More expensive if separate equipment is provided to each student.
2. Difficult to schedule in the school timetable when double periods are provided.
3. More time consuming compared to demonstration method as students are unskilled workers and or not as competent to handle apparatus as their teachers.
4. No guarantee that the students will learn to solve problems or think scientifically.

When there are more students and the equipment is not enough for each student to do experiments, then a large number of experiments should be arranged, and different students could do different experiments at the same time. If possible typed instructions should be given to students for each experiment. This will facilitate students to work on their own and teacher will have time to go to students who need him most.

## Enquiry Project Methodology

The Investigatory Project Method involves investigation, discovery and finding out something which was not known to the student before. An investigation is much more than the repetition of a standard experiment. Here the student is to decide what experiments are necessary and how he is going to carry them out. He may have to design his own apparatus, if that is not available in the laboratory. He has to search for the appropriate principles, laws, formulae, apparatus and data; and originate solution to a problem. The student will act like a scientist.

This method involves the steps of scientific method:

1. Problem;
2. Hypotheses;
3. Experiment
   (a) Testing of hypotheses by experiments
   (b) Collection of data
   (c) Interpretation of data
4. Conclusion.

This is the least used method of teaching science in our schools. In USA, Canada and UK teachers try to cover a part of syllabus by this method. In this method a group of students select a problem in consultation with their teacher/ and then through discussion develop the plan of action and design the type of equipment needed to carry out the experiments to reach the conclusion. Students thus are trained in science processes and scientific method.

***Advantages:***

1. It creates interest in science.
2. It develops understanding.
3. It develops self-confidence, cooperation, leadership and emotional stability.
4. It promotes curiosity and develops scientific temper, interest and appreciation.
5. It develops abstract and concrete scientific skills.
6. It develops interest in scientific hobbies in the right use of leisure time later on.

***Limitations:***

1. This is the most difficult method for the teacher, requiring more planning and more effort of execution.

   Though it is the most difficult method for the teacher requiring more effort and planning, the lack of this effort is one of the reasons for much of the mediocre teaching that is done. The results of this method, when it is properly executed, are more than compensate for the extra work that is required.

2. If properly operated it calls for somewhat more materials and equipment than in the strictly demonstration method.

   The necessity for more materials and equipment is generally an excuse rather than a reason for not using it. The problems and experiments can be so devised that they will require nothing that cannot be had at little or no expense.

3. The amount of time consumed by the projects may be so great as to make it impossible to cover a wide variety of topics thus resulting in an incomplete course.

   Also to the economy of time, there is much doubt as to whether time spent upon less subject matter might not be worthwhile to the student than the same time spent in hurrying over more subject matter.

4. It may result in a task of proper coordination of subject matter when individualised, since students may be working upon different problems or projects.

   The lack of coordination which sometimes is the result of individual work is little more than the result of poor planning and less execution upon the part of the teacher.

5. Difficulties of method increase when there are more students in large classes and more periods per day of the teacher.

   In no other method there is equal provision for the essential factors demanded by our objectives. The solving of problems is essential to the training in scientific thinking and the following of projects is a vital part of the production of those attitudes which should be the outgrowth of science teaching. Proper planning will provide for group or even class participation as a means of diminishing the problem of number of students and number of classes that must be taught and will still make possible a sufficient amount of independent work to provide for the necessary individualisation.

## Solution Giving Mode

Problem is not a problem if a child can solve it by his previous knowledge. Problem solving takes place only when student's previous knowledge is insufficient to enable him to provide an acceptable solution, and solution becomes possible only when he acquires new knowledge which he does not have before. A method to solve a problem is called a problem solving method.

***Steps in Problem Solving Method***

1. ***Problem***
   - (a) Identification of problem after problem survey;
   - (b) Statement of the problem — a clear description;
   - (c) Explanation of the problem — discussing with students ensuring that they clearly understand what the problem is;
   - (d) Delimitation of the problem — isolating only those parts of the problem which are within the reach of students.
2. ***Hypotheses***
3. ***Experiment***
4. ***Concluding***

These are the four steps of scientific method. Scientific method is a problem solving method.

In problem solving method step III may be eliminated, and for testing the hypotheses information from some other sources (reference books, journals, newspapers, discussion with some experts, etc.) may be taken.

Sometime even steps II and III both may be eliminated in problem solving method, and information from some other sources (mentioned above) may be used to find the solution of the problem.

*Advantages*: Its advantages are the same as of Scientific Method or Project Method of teaching science.

*Disadvantages or Limitations*: Its limitations are also the same as of Scientific Method or Project Method of teaching science.

## Personal Guidance Mode

Not all students are alike, but we teach them the same thing by the same method. In a classroom there are students of different science background. Most of them perhaps do not have aptitude for science, if they have a choice perhaps they will not offer science. But we are to teach the same science to all of them, and we usually use the same method to teach all students, without taking into consideration their individual differences.

Usually we can classify our students in three groups — average, below average and above average. When we teach science we assume some pre-requisites or entry behaviours of our students, we assume that this much they already know, and then we start our lesson. When we do so, we take into consideration only the average students. The result is that only the average students are satisfied and the below and the above average are frustrated. The only solution remains that some individualised instruction and some other self-learning materials be developed and used to satisfy all the students.

Programmed Instruction is one way of individualised instruction. Self-learning modules, slide-tape programmes, and multimedia packages are some other means. Though such materials are not available in the market, yet they can be developed by some innovative teachers for the use of their students. Then their materials can also be shared by other science teachers in their classes.

## Modular Method

Use of self-learning modules in teaching is another form of individualised instruction. This is called Modular Approach of teaching and learning. If self-learning modules are available on some topics, they can be given to students as assignment for self-learning. Teachers can also develop such modules on some topics. Such modules once developed can be used again and again.

## Student and Teacher Centred Method

### *Teacher Centred Approach (TCA)*

If we observe classrooms, we find it is the teacher who speaks most of the time, and the children either given no opportunity to speak or a very little opportunity to speak — ask questions or interact. This is a Teacher Centred Approach of Teaching.

***Example:*** In a study about 1000 classes were observed (I-XII) in all subjects and in all types of schools, and it was found that on an average 95 per cent of the time teacher was speaking and 5 per cent of the time the children were speaking. This is an example of Teacher Centred Approach of Teaching.

Lecture Method and Lecture Demonstration Method are Teacher Centred Approaches of Teaching. Teacher Centred Approach is a very effective method for teaching skills (processes). Reading a thermometer, reading a spring balance, preparing gases like hydrogen, oxygen and carbon dioxide, salt analysis, volumetric and gravimetric exercises, setting up a microscope, making a slide and seeing it through a microscope, setting the experiment for photosynthesis, etc., are all skills.

Until a child perfectly learns a skill, the teacher is very much involved in telliing the child "what to do" and "how to do." The teacher has to use Teacher Centred Approach. Cooking; sewing; driving a scooter, a motor cycle or a car; repairing; playing musical instruments; dancing, etc., are all skills. Teaching each of these skills, a teacher is needed who teaches these skills by Teacher Centred Approach.

***Example***: Reading a thermometer is a skill. According to a study children and even teachers not taught this skill by Teacher Centred Approach, generally read the thermometer wrong. If the reading was 46°C, they read 40.6°C. Then came the teacher to teach this skill. He pointed at 40 and asked "What is the reading?" The child said "40°C". Then he pointed at 50°C and asked "What is the reading?" The child said, "50°C". Then the teacher asked, "How many marks are there between 40°C and 50°C?" When the child failed to answer, the teacher himself counted — 1, 2, 3, 4, 5, 6, 7, 8, 9, 10.

Then the teacher asked "How much is one mark?" Then the child replied "1°C." (This is actually the least count of the thermometer which the child was not knowing, and therefore he read 40.6°C instead of 46°C). Then the teacher asked, "How many marks is the mercury above 40°C"? And the child counted, "1, 2, 3, 4, 5, 6." Then the teacher asked, "Now tell, what is the reading of the thermometer?" And the child again counted, "41, 42, 43, 44, 45, 46," and then said," Thermometer reads 46°C and not 40.6°C".

Now look at the approach, teacher used to teach the child to read the thermometer correctly. In this approach the teacher was more involved than the child, and therefore this is Teacher Centred Approach. Reading a thermometer is a skill. Therefore, Teacher Centred Approach (TCA) should be used to teach skills (processes) to the child.

### *Child Centred Approach*

Some innovative teachers themselves speak less in the class, and give opportunity to their children to speak more. Here the science teacher acts as moderator (leading and moderating discussions) and a guide (helping slow learners).

This is Child Centred Approach. In teaching science by Lecture Method or by Lecture-Demonstration Method, these innovative teachers involve their children in interaction and question-answer sessions that even their lecture and Lecture-demonstration become child centred. Enquiry Approach, Problem Solving Approach, Laboratory Method, Scientific Method, Project Method and Investigatory Project Method are Child Centred Approaches.

***Example***: A teacher is teaching a concept "Air has weight" by scientific method. It involves *"weighing"* by a spring balance, which is a *skill* (*process*). The teacher teaches the skill (process) of weighing to the children by Teacher Centred Approach before he teaches the concept "Air has weight."

Then he starts the lesson with a:

1. *Problem:* Does air has weight? (This is TCA). Then children formulate their own.
2. *Hypotheses:* (i) Yes, (ii) No. (This is CCA). Then children (as already know the skill of weighing or reading a spring balance) test their hypotheses themselves with an experiment.
3. *Experiment:* weighing an empty football bladder, and the same football bladder full of air. (This is CCA). Then they themselves draw the conclusion.
4. *Conclusion:* that air has weight. Hypotheses "yes" is right, and hypotheses *"no"* is wrong. (This is CCA).

Here step — 1 is Teacher Centred as teacher asks the question or poses a problem. Steps — *2, 3, 4* are Child Centred. Therefore, Scientific Method is 25 per cent teacher centred and 75 per cent child centred. Thus, as a whole Scientific Method is Child Centred Approach (CCA).

Note "weight" is a concept and "weighing" is a skill. When skill of "weighing" already taught by the teacher to the children by Teacher Centred Approach, the concept of "weight" is to be taught by Child Centred Approach. Do not tell to the child "air has weight," let him find out himself whether or not "air has weight" by the Child Centred Approach (CCA).

***Remember:***

1. Teach skills (processes) by TCA, and
2. Teach concepts by CCA.

## Heuristic Approach

Heuristic means discovery in Greek. Professor Henry Edward Armstrong of Imperial College, London, was a strong advocate of Heuristic Method. Professor Armstrong tried his best to place his students in the position of original discoverers. He involved his students in finding out by themselves instead of telling them. Laboratory work was compulsory for every student where they could discover instead of being told. His method was quite simple. He used to provide instruction sheet to his students concerning the problem and students were expected to take observations and conduct experiments according to these instructions. Students used to record their observations in their record-books. From these observations they were asked to draw their own conclusions. Thus, they were introduced to reasoning from their own observations and experiments.

Heuristic Method demands that the students should be allowed to stop and think, discuss and suggest modifications for further experiments.

*Advantages and Disadvantages*: Heuristic Method develops spirit of enquiry in students, but this method can never be the main method for science teaching. This method is a time consuming process by which students gain less worthwhile and significant knowledge in more time. Students are not expected to rediscover all the knowledge which already exists; they are expected to know at least a part of it.

## Enquiry Method

***Example***: One day an electric bulb blew out in the class. "What happened" asked the students. The teacher took out the bulb from the bulb holder, and showed it to the students. The students gathered around the teacher. He passed it around them and said, "Look at it and try to develop a hypothesis about what happened." "What is inside the bulb" asked one of the students. "I do not know," said the teacher. "Is there air inside the bulb," asked another student. "No," said the teacher. "Is there any other gas inside the bulb," asked another student. "No," said the teacher.

The students were puzzled and they started looking at one another. Finally one student asked, "Is it vacuum inside the bulb"? "Yes," said the teacher. "Is it complete vacuum," someone asked. "Almost" replied the teacher. "What is that little wire made of," asked another student. "I do not know," said the teacher. "Is that little wire made of some metal," asked some students. "Yes," the teacher said.

Such Yes-No Answer questions continued between teacher-students interaction, till students identified the material of the wire inside the electric bulb, and the events that took place without teacher's readymade answers. Finally students began to formulate hypotheses about what happened. After the students formulated some hypotheses, they started searching through Reference Books (in the library or at home) in order to verify them.

This is an example of enquiry Approach of teaching science. Students are trained for enquiry. Enquiry begins with a puzzling event like "an electric bulb blew out in the class." Students enquired (what happened — a problem) when they were puzzled.

After the puzzling event is presented to the students, they ask the teacher some questions. The teacher should answer the questions in "Yes" or "No." The answer of each question may be a small hypothesis.

Inquiry Approach of teaching science may have the following steps:

1. Encounter with the Problem.
2. Formulation of Hypotheses.

3. Data Gathering — Experimentation and Verification.
4. Conclusion – Solution of the problem.

## Debating and Clarification Mode

When teaching science by lecture method or lecture-demonstration method, if students are involved in discussion (interaction) they are in a much more better situation of learning science. As discussed under lecture method, the teacher should not lecture continuously for more than a very limited time so that the teacher should not lose the concentration of their students.

For this he can do a change of activity, and discussion (interaction) is one of the several activities in which students can be involved very actively. So by using lecture-cum-discussion method some of the disadvantages of the lecture method can be taken care of.

Similarly when lecture-demonstration method is combined with a well directed discussion (interaction), it may prove to be a successful technique of teaching science.

In our schools lecture method and lecture-demonstration method are most commonly used. If we add "discussion" or "interaction" into it, these methods may come out to be much more effective,

There may be 4 types of interactions:

1. Teacher-Learner Interaction (TLI)
2. Learner-Learner Interaction (LLI)
3. Teacher-Material Interaction (TMI)
4. Learner-Material Interaction (LMI)

When teacher is lecturing, TLI may be initiated. When children are discussing in small groups, like tutorials, LLI may be initiated. When teacher is demonstrating, he is using some teaching aids (material) he can use TMI along with TLI for more effective teaching. When children are doing some experiments in small groups, LMI will be of better use along with LLI and TLI.

## Programmed Guidance Approach

Programmed Instructional Materials as discussed in the elsewhere, provide self-learning units. Such units may prove to be very useful when our science teachers are heavily loaded, and they do not have time to interact specially those students individually, who really need them.

There are some topics in every field of science (Physics, Chemistry and Biology) which the students can learn on their own if programmed learning units are available on these topics. But in our country programmed learning units are not available in the market. So the only choice remains that some innovative science teachers should write some programmed units on selected topics to be used by their students. These guidelines may be used when writing programmed units. Such units when written, may be shared by other science teachers in their classes. Self-Learning Modules are as good as Programmed Instruction Materials for self-learning.

## Method of Multimedia Package

Like Programmed Instruction, Self-learning modules and Multimedia Packages if available on some topics, may also be given to students to cover part of syllabus on their own. Multimedia packages can also be developed by teachers. Such packages once developed may be used again and again. This is Multimedia Package Approach of teaching.

## Teachers and Science Education

"Teacher! I saw a satellite!" "Look! One baby mouse is white!" "Can we attempt the experiment this way?" "I have collected all the data — use of log tables and even slide rule is too time consuming and out of date for calculations. May I use a calculator? Please! Now we have computers."

Everyday, they are different — but everyday greetings like these begin the day for science teachers. One day it is satellite, one day the moon, one day how the seesaw operates, one day the why of earthquakes — or another question everyone is asking. Science teachers as well as students; government as well as commentators — everyone needs the answers to questions in science.

It is a wonderful world when you look at it this way — and science is like a key that opens doors everywhere — on the bottom of the ocean — with the astronauts in space — with the plants, animals, and man here on earth — with thousands of conveniences of our daily lives. Learning about science — as you already know — stretches your mind, but gives you a lot in return. You always have something new to bring to your classes. And you as a science teacher have many satisfactions that come from working with young students.

You have been a practising science teacher and now you are ready to be regular science teacher, teaching a class of physics or chemistry, biology or integrated science, you will surely like to be a good science teacher. This is the right time to ask yourself some questions like these:

- What does it take to be a good science teacher?
- How can I use my teacher's training to be a good science teacher?
- Where can I get some more information about good science teaching?
- What things should I do to begin with to be a good science teacher?
- What should I do to be happy as a science teacher?

## Preparations by a Good Teacher

After completing your teacher training programme, this is the right time to ask yourself whether you possess the basic intellectual and personal qualities necessary to be a good science teacher. Do you have the attributes, such as integrity, drive, and a high sense of responsibility, which are valued in teaching specially science teaching. Besides these, there are special characteristics of successful and happy science teacher.

Checking the following questions may help you decide, how good a science teacher you are going to be.

| | *Yes* | *No* |
|---|---|---|
| 1. Do you enjoy reading and studying in science as well as in other academic areas? | ..... | ..... |
| 2. Have you been a good science teacher? | ..... | |
| 3. Do you like working with students and other science teachers? | ..... | ..... |
| 4. Do you like to help your students? | ..... | ..... |
| 5. Do your students and other science teachers like to work with you? | ..... | ..... |
| 6. Could you enjoy and adopt to a career in which the subject is constantly and rapidly changing because of new discoveries? | ..... | ..... |
| 7. Do you/ on your own, sometimes do more work than is required by teachers? | ..... | ..... |
| 8. Have you developed science projects and participated in science fairs? | ..... | ..... |
| 9. Do you like to find answers to problems on your own? | ..... | ..... |
| 10. Do you have a sense of humour? | ..... | ..... |
| 11. Can you accept criticism and profit from it? | ..... | ..... |
| 12. Do you consider that preparing students for living in the World of the future is a challenging job? | ..... | ..... |
| 13. Are you developing the ability to read quickly and with comprehension and to express yourself clearly and interestingly in speaking and writing? | ..... | ..... |
| 14. Are you in good health, both physically and mentally? | ..... | ... |

If you have checked thoughtfully YES for most of the above questions, you very likely would enjoy teaching science at any level, and you should read on to learn how you prepare for this type of exciting career.

## Tasks of Science Teacher

Everyone needs an education in science in order to understand the world and to be a good citizen and good worker. Scientists, leaders in government, industrialists, and the public in general need to be able to understand scientific problems and to make good judgements about them.

We need scientific literate and informed people to make decisions about such things as major government projects in space exploration and medical science; health problems, environmental and population problems, such as air and water pollution; diseases such as cancer or tuberculosis and epidemics; or the possible effects of chemical pesticides upon us and our environment.

More and more people will need science in their work, and they must depend on teachers to help them get the knowledge they will need. Each new "breakthrough" or the expanding frontier of science and technology, and the public's need to be informed, require a never-ending supply of well-trained people to fill the ranks. This is also true in engineering, medicine and other fields related to science.

The next generation of youngsters, as it passes through school, must be as well-educated in and about science as its talent will permit if we are to continue to prosper in our democratic society.

Moreover, the ever-increasing scientific and technological progress challenges man's ingenuity to improve his methods of processing, storing, retrieving and reporting all sorts of scientific information. In a sense, we need "new breeds" of people who can combine science with other talents, as for example, the scientist, librarian and the science reporter.

Good science teaching is one of the most valuable ways to meet this urgent need for science-educated citizens and workers. Enthusiastic, intelligent, and well-educated science teachers inspire and prepare students to investigate the great questions of science and the questions raised by the scientific discoveries which affect us and our society.

Mainly through the inspiration of devoted science teachers, great number of students develop lifelong scientific interests and learn to appreciate and understand the nature of science and its usefulness to mankind.

The science teacher derives great satisfaction from this special importance of his work. And added to it is the great satisfaction that comes from mastering a field of scientific study and from affording a special kind of service to others.

So, when you ask yourself: "What do science teachers do?" You may answer:

"Everyone today needs to understand what science is about. We need it because discoveries in science affect many aspects of society. To be responsible and useful citizens, we must be able to judge how the discoveries of science may be best used for ourselves and society. Many among us will make a career of science, engineering, or technology. We need a good foundation in science to become competent in our chosen work. Therefore, we can say in general that a science teacher, at whatever level he is teaching, is laying the foundations for an understanding of science."

A science teacher is a teacher first and then a science teacher. As a teacher he should be understanding, sympathetic. Teachable and free of prejudices. Science teaching requires a sound knowledge of the subject and a real interest and ability in sharing this knowledge with others. It also requires keeping up-to-date. The work is not easy, but it is creative, and it can be extremely exciting and satisfying. It takes thought, energy, and enthusiasm.

Here is an example of what one good school expects of its science teachers:

1. He plans his work for the whole year before school starts its session.
2. He plans his lessons well in advance before he enters the classroom for teaching science.
3. He knows the various practical skills needed in his particular field.
4. He guides and assists his students in their laboratory experiments, field trips and projects. He is also responsible for the safety of the students and the condition of the laboratory and its equipment. He decides in what way each aspect of his subject can best be taught. Some of the ideas and problems can be most effectively presented by demonstration, others through student investigation, field trips, or by films or lectures. He also masters the finer art of asking the questions that will stimulate his students to think for themselves and to search for answers.
5. He organises class procedures, plans the use of the laboratory, discusses with the class, and helps his students learn how to find information outside the classroom in school and city libraries, local industries, or by consulting experts in the community.
6. He encourages his students to develop a lasting interest in science, its changes, and its methods. He guides both individual and group projects and help the students participate in various programmes of awards and recognition. He may sponsor a school science club and with the help of other science teachers and students, organises a science fair or plan field trips.
7. He keeps both himself and his students informed on the most recent developments in his field. He draws his students attention to the social, economic, political, and other aspects of the relationship between society, people and modern science. Thus, he and his students always see that science study is not limited to the classroom.

8. He attends orientation courses in his field (seminars, workshops, summer institutes), as well as science conventions and science conferences for his academic growth.
9. He participates in science curriculum development programme conducted by various agencies like DIETs, SCERTs, LASEs and NCERT, etc.
10. He conducts action research in science education and actively participates in innovative science programmes for better science teaching.
11. He always efforts for quality science education to his students. Now you can figure out yourself what should be the competencies for a good science teacher (Physics/Chemistry/Biology).

4

# Science Teaching Objects

The aims of teaching science should be formulated in the form of attainable objectives. It is no use setting major goals of teaching a subject, just to embellish the pages of the book. Objectives to be of practical value should be worthwhile and attainable, and should be stated in unambiguous and specific terms. But before formulating the objectives, a study of the needs of the individual in his society is essential. These objectives should help and guide us to select and organise learning materials. While selecting objectives, we must take into account the age and the interests of the learners and also keep in view the behavioural change we want to bring about in them. That's the basis of objectivity.

## Teaching Objectives

Teaching of any subject has broader objectives. The aims of teaching any school subject must always be directed towards achieving the aims of education in general. We must remember that whatever subject is taught to the learners, the ultimate aim is to educate the child through the teaching of that subject. Thus, in science education too, we should impart instruction in the subject not so much for the sake of the subject but to serve as an efficient agent or medium for properly educating the child. The aim of education has been defined in different terms by different authorities. But whatever be the diversities in the literature of the definitions, there is no controversy over the common expectation that education must develop to the full, the potentialities of the individual learner in accordance with the general good of the society of which he is a member. The students may also try improvisation of different types of instruments. Such activities provide opportunity for utilisation of leisure and fulfil one of the aims of education, and soon.

The teaching of science as a subject must, therefore, contribute to the all-round development of the child so that he comes out as a socially useful and efficient citizen of the modern scientific world. This aim is, however, too general and too broad to be meaningful in the classroom situation. The broad aims should serve as a frame of reference and should be regarded as directions of growth. It is, therefore, necessary to formulate suitable objectives to serve as roads towards attainment of the ultimate ends. These objectives should be such that the learners' growth and development are directed towards their realisation and they are practically attainable through the teaching of the subject.

The formulation of practically achievable objectives of science teaching brings into consideration a number of factors that are related to the learner and the process of learning. Therefore, while selecting suitable objectives of teaching the subject, we must keep in view the psychology of learning, principles of teaching and the stage of the learner. The formulators of the objectives must take into account not only the need and the abilities of the learner, and the character of the subject but also the demands and expectations of the society. Further, the objectives should be so selected that the progress towards their realisation could be apprised and evaluated.

Practically attainable objectives can be formulated not only for teaching of the subject as a whole but also for each topic and each lesson. Keeping in view the general objectives of science teaching specific objectives can be framed which may be attainable in the classroom situation. There should be a simple series of specific objectives which can be used by the science teacher in his daily work to give him direction and guidance. Thus, the specific objectives are those which can be measured in the classroom.

These refer to learning, skills and interests which can actually be developed through classroom teaching. The recent trend in science teaching has shown a distinct tendency of general change in emphasis from mere acquisition of assorted facts to the development of functional learning, from unrelated knowledge units to contents logically organised about problems that concern the pupils and from preparation of specialists to general education in science from all. But it does not mean that the future specialists will remain uncared for.

## Science Learning

The schools must give the future specialists a good foundation for later learning. The modern world demands that every citizen know about science to be an efficient member in the society. It is the duty of the school to maintain a balance and the two jobs must be carried forward simultaneously. But the standard of knowledge and the speed of teaching need not be the same for all. The gifted pupils will be identified and nurtured by giving them opportunities to progress at their own speed.

It is the school to which the nation will always look for the future scientists and technologists or specialists in the other fields of science. About the general education in science to all, the fifty-ninth Year-Book of the National Society for Science Education says that all persons need to know about science "not so much as a great wealth of knowledge as a happy understanding of the nature of science. What everybody needs is a sense of knowledge leading to a sympathy with science and an appreciation of the way the scientists work." Thus, the duty of the school is not only to help the pupils acquire the requisite scientific knowledge but also enable them to appreciate the part science plays in their lives.

The school will, during the process of teaching, inculcate the higher virtues of science teaching such as critical thinking, accurate observation, scientific method and scientific attitude. The school cannot evade the obligation of making the pupils aware of the social aim of teaching science.

The pupils must be given sufficient understanding of the facts of good and healthy living in the society. They should be explained how the knowledge of science can be utilised for the benefit of the individual and the society. Since schools prepare the future citizens, it is their duty to equip the pupils properly for their future life. In this modern world, the preparation for life is incomplete without the knowledge of basic scientific facts and principles and their applications in everyday life. Though the schools must give the future specialists a grounding in their fields of interest, it is not the duty of the schools to prepare specialists.

The schools must give more importance to teaching science as a part of liberal education to pupils, since they prepare them for living in this scientific world. Every teacher who has to teach science in schools, will realise that whatever aims or objectives we formulate, will remain as paper goals, unless the teachers themselves make conscious efforts to achieve them through their teaching. For this purpose, they should employ the modern methods of teaching science, take help of the available aids for teaching science, evolve creative activities for the pupils and try to accomplish the ends as far as practicable under the circumstances in which they work. The primary objective of science teaching is to give knowledge and information about the world we live in.

To live as an efficient member in the modern society, each citizen needs to know some facts of the natural phenomena (living and non-living), laws and properties of matter, and the application of the knowledge of science and the scientific principles that we come across in our daily life. The school science course should therefore be a part of sound liberal education and thus be general, based on humanistic lines so that it may be of value to all. But such body of knowledge or information must be related to and based upon their familiar experiences in the environment.

Every pupil must acquire an adequate background of ordered knowledge that is useful in maintaining health, promoting safety, and in understanding the environment. The recent theories and new knowledge is significant and should be included in the science course. This, however, immediately raises the question of selection from the mass of ever increasing scientific knowledge. One of the objectives of science teaching is to make the pupils understand science in its various aspects. They should be acquainted with the different branches of science which will enable them to understand and interpret their environment.

The students should also be made aware of the successes of men in using and exploiting nature. They should know about the things, materials and powers harnessed in nature—the Sun, water, forests, mines, etc.—for the benefit of man. They should also learn about the character of scientific knowledge, how it is developed and used. In short, the science teacher should help and encourage pupils to acquire an interesting and useful acquaintance with all aspects of science, important in the modern world.

In a democratic society each individual is responsible in shaping the public policy of a country. Thus every one needs to understand the relation between science and men, and the interplay of scientific and technological innovations and human affairs. Every young man and woman should be able to understand the dependence of any modern society upon the scientific inventions and to realise that science is an integral part of our modern living. Science is not just a collection of some isolated and assorted facts and principles. Moreover, knowledge is not static. It may change its meaning and importance with time.

Science, to be meaningful and valuable, should be taught as a whole, at the school stage. The pupils should find themselves in a world of related happenings. The facts and principles of science should, as far as practicable, be taught as generalised concepts.

There are many facts or events which have fundamental similarities and a common basic explanation. In this connection H.N. Saunders' *The Teaching of General Science in Tropical Secondary Schools* states that if science is taught in this manner, the pupils will discover that "vibrations and movements tend to travel in waves, that living cells everywhere conform to the same pattern, and that every flower the world over, is a variant of even if it does not achieve, a standard arrangement.... A pupil may learn of several extraordinary different instances of resonance, but his amazement at their variety must be supplemented by appreciation of the fact that one explanation is common to all: should understand and be stimulated by, the idea of that one single thread woven through the whole fabric." The Committee on Science Teaching of the National Society for Study of Education (Thirty-first Year-Book, Part I) even goes to identify the specific objectives of science teaching with the generalisations or the principles such as:

- All life comes from life and produces its own kind of organisms.
- All life has evolved from simple forms.
- All matter is probably electrical in nature.
- Chemical changes are accompanied by energy change.
- Radiant energy travels in straight lines through a uniform medium.
- The energy of solar radiation is continually working changes on the surface of the earth, etc.

All such generalisations or concepts may not be comprehensible to all grade levels.

The teachers should plan simple experiences for the pupils keeping these broad concepts in mind. Thus one of the objectives of science teaching is to lead the pupils and acquaint them with the broader concepts, generalisations and outlooks. The most important objective of school science instruction is to make the pupils aware of the scientific methods of procedure and to inculcate scientific attitude of mind.

The school will not only give the pupils adequate scientific knowledge and requisite skill to meet the problem of existence but also train them in proper scientific methods of investigating problems and create a scientific attitude in them. The methods of investigation are as important as the facts themselves. About the importance of training the pupils in scientific method, the Science Masters Association of England reports: "This method should become clear through our teaching, no lessons are given directly on the meaning of scientific method, rather does it become evident from the spirit of our teaching. The method should become a technique—in fact, an attitude towards life."

The pupils should be asked to devise and perform experiments themselves, observe carefully and accurately the happenings, collect relevant data free from any bias or prejudice, organise the collected facts systematically, think critically before forming their conclusions. The science teacher should always encourage an attitude of discovery. Such an independent and impartial experimentation helps the pupils to develop a logical mind, critical judgement and the habit of solving problems independently to find answers to their own questions.

The independent way of solving problems will be of great help in their later life. This, however, does not mean that we are attempting to produce science specialists. The scientific way of handling any problem and a scientific attitude of mind should be inculcated in all individuals in order that they do not accept things on hearsay, propaganda or superstitious traditions but upon conclusions arrived at on the basis of evidence. Even an educator like John Dewey has emphasised method and attitude as objectives of formal education. This, in fact, is the greatest contribution of science teaching.

## Systematic Process

The scientific method is sometimes defined as the systematic process or procedure, where the aim of doing something is combined with the aim of understanding what is done. It should be borne in mind that a mastery of facts, principles and concepts of science will not develop in the learner the desirable outcomes of science-learning such as scientific attitudes, or scientific methods of procedure. The science teacher must deliberately develop these qualities through the process of instruction in the classroom.

If the teacher is sincere and tactful, he can, by providing proper science situation, lead the pupils to develop sensitive curiosity to discover the problem to be investigated, to form the habit of seeking evidence before forming conclusions, to accurate observation, to the habit of being patient and to withhold until sufficient evidence is available, to develop the ability to select relevant data, organise them systematically and form conclusions free from bias, prejudice or hearsay.

The scientific attitude also includes open-mindedness and respect for others' opinion, but the latter does not mean that the pupils must be so credulous as to take things for granted without adequate grounds. The impact of modern science on society has raised the problem of acquainting the pupils with the social implications of science teaching. The students of the modern world need to understand and appreciate the dependence of a modern society on science and the changes in the social structure that have been brought about by the achievements of science and technology. They should not only be able to appreciate and wonder at the modern marvels of science but should also understand the social use of the scientific achievements.

This can be justified from the point of view of the fact that modern liberal education has a much wider orientation, and thus, this idea of developing social attitude and appreciation, should be considered as an aim of teaching science. The science teacher should teach science in such a way that the pupils understand the social functions of science, think and act in relation to the implications of science and the society. They should appreciate science as a part of modern living and that science should always be used for the benefit of the society. While learning about the achievements of science, they should be made to realise that the scientific achievements should never be used for the benefit of an individual at the expense of the society.

Developing a social aim or the social attitude and appreciation seem to be growing in importance in the present time as a world opinion, since it is seen that the scientific achievements have been directed towards unfortunate ends. It should be the duty of the schools to train the new generations, who are going to be future citizens, to utilise knowledge of science for social good. A study of the lives of the discoverers helps to appreciate their discoveries better and makes the facts more interesting and lively.

It will, be worthwhile to give the pupils brief biographical account of the lives of the great scientists and discoverers, their attempts to conquer and exploit nature, their attitudes and their sacrifices for the good of the society. They should be made aware of the fact that science is now no longer the job of a single man and that success, in most cases, results from a cooperative search of many; they should also be able to appreciate the difficulty of original thinking and to apprehend how superstitions and wrong beliefs were dispelled by the efforts of the great scientific geniuses like Copernicus Galilio, and others. What the pupils learn will not be of lasting value if they do not take interest in it. Science materials should be presented in an interesting way so that the pupils are motivated to learn more and more about science. They should be led to find an engrossing interest and joy in studying science.

**Awakening of Natural Interest**

The teacher of science should aim at awakening pupils' natural interest in learning science and about its multitudes of marvellous manifestations that we see everywhere around us. A science programme can be made interesting by organising science club activities, science fairs, visits to places of scientific interest, field trips and excursions. In the highly industrialised urban areas, it becomes a problem to utilise the leisure time. Such a state creates the need for opening new avenues of interest to use the leisure hours.

Scientific activities provide satisfactory hobbies for this purpose. Moreover, such activities and hobbies give enjoyment and satisfaction which help normal development of the pupils. These creative activities satisfy the psychological needs of the growing boys and girls, and provide healthy pastimes not only to the school-going generations but also to those who have left school and have enough leisure time. The science teachers should realise the importance of this aim of teaching science. Various commissions, committees, associations and other academic bodies have defined the aims of science teaching. All these reports contain objectives covering areas of knowledge, skills, abilities, habits, interests and appreciation as well as a basis of specialisation and training in better living.

The Secondary Education Commission states: "Science in secondary schools is not directed to the production of scientists. Its aim is to give basic understanding and an appreciation of scientific phenomena, biological and physical, which may prepare the 'non-scientists' for a fuller and more complete life. At the same time the course should give fundamental principles to those relatively few who will specialise in science.

The teacher should aim at awakening in the pupils a lively curiosity about the natural phenomena arousing and developing their capacity for the practical application of their knowledge, at appreciating the tremendous impact of modern science on all aspects of our life, at interesting them in the human side of all the scientific progress by introducing them to the lives of great scientists. Such an approach will ensure that science becomes a part of 'liberal' education and an instrument for the appreciation of special characteristics of modern culture." The 'All India Seminar on the Teaching of Science in Secondary Schools' (1956), has stated the aims of teaching general science in Secondary Schools as:

- To familiarise the pupil with the world in which he lives and to make him understand the impact of science on society so as to enable him to adjust himself to his environment.
- To acquaint him with the 'scientific method' and to enable him to develop the 'scientific attitude'.
- To give the pupil a historical perspective so that he may understand the evolution of scientific development.

Further, the Directorate of Extension Programmes for Secondary Education, Government of India, in its brochure on 'Evaluation in General Science' sets some of the objectives of teaching general science in secondary schools as follows:

- The pupils studying general science should acquire knowledge of the fundamentals of science useful to all in everyday life.
- They should develop the ability to apply the knowledge in everyday life.
- They should acquire experimental skills such as: (a) handling apparatus and instruments; (b) arranging apparatus for an experiment; and (c) preserving apparatus, chemicals, specimens, models, etc.
- They should acquire constructional skills such as: (a) improving simple instruments and appliances; and (b) repairing certain instruments and appliances of everyday use.
- They should develop drawing skills such as: (a) drawing and sketching certain objects, instruments and arrangements; and (b) photography in certain objects and specimens.
- They should be able to locate reliable and recent information from appropriate sources.

- They should be able to interpret scientific data given in various forms such as tabular, graphical, scientific, etc.
- They should develop the power of minute observation of their surroundings.
- They should develop the power of oral expression in science to discuss, argue, describe and raise questions, using scientific terminology.
- They should develop the scientific method in thinking and action.
- They should adopt the scientific attitude in making statements, accepting information and forming beliefs.
- They should develop interest in scientific reading and hobbies.
- They should be able to appreciate the impact of science on life, both personal and social, the struggle through which science has advanced, and the inspiring work of the scientists.

Further, the following similar set of objectives was formulated by the principals of Delhi Higher Secondary Schools in the third summer camp organised by the Extension Department of the Central Institute of Education, Delhi;

- To develop in the student a scientific attitude.
- To develop in the student critical thinking.
- To enable the student to acquire the fundamentals of scientific method.
- To enable the student to be creative.
- To develop in the student skill in laboratory techniques.
- To develop in the student the ability to apply scientific knowledge and principles to problems of everyday life and new situations.
- To enable the student to comprehend scientific terms, concepts, symbols, various tables and their uses.

- To enable the student to construct and interpret graphs, diagrams and models.
- To enable the student to collect and interpret data for the solution of problems.
- To enable the student to be familiar with the natural resources of his environment and their uses.
- To enable the student to be familiar with the trends in modern science.
- To enable the student to appreciate the beauty and order in nature.

Later, the "All India Seminar on Teaching of Science in Secondary Schools" organised by All India Council for Secondary Education, at Taradevi (1956) gave details of the aims of science teaching at different stages of education. These may be summarised as follows:

- At the primary level of school education, the aims of science teaching should be to arouse and sustain the learners' interest in nature and in their physical and social environment. Through learning science, the pupils should develop love for nature and observe the things in nature. Science teaching should inculcate the habits of healthy living, cleanliness and orderly behaviour. It should arouse curiosity and develop in the pupils habit of observation, exploration and systematic thinking. Science learning should also develop their manipulative and creative abilities.
- At the middle level, in addition to the above, the pupils should acquire clear information of nature and its events so that this knowledge serves as a basis for learning general science course at the secondary stage. They should develop the ability for generalisation and an understanding of the application of science in everyday life. Science teaching at this stage should aim at inspiring the pupils through the stories of the lives and achievements of the scientists. They should be able to see

the impact of science on the individual and social living and also develop interest in pursuing scientific hobbies.

- At the high and higher secondary stage, the pupils should be able to understand and realise the impact of science on society and also learn to adjust to the environment. The most important aim is that in addition to acquiring knowledge of science, they should develop a scientific attitude in them and receive training in scientific method. At this stage, they should be familiar with the historical perspective of the evolution of science and its later developments.

After that, the NCERT suggested the objectives of teaching general science at the elementary stage of education covering the age group 6-14 as follows (General Science Syllabus, NCERT, 1963):

- To acquire knowledge of biological, physical and material environment.
- To develop skills of solving problems and their application to life problems, to locate problems as well as to design procedures to solve them, collect proper data and methodically organise them, observe accurately and interpret results logically.
- Develop scientific attitude of mind and inculcate good personal and social habits such as objective and unbiased outlook, love for truth, inquisitiveness, accuracy and precision, right health habits, habit of enquiry, initiative and logical thinking.

4. Develop interest and appreciation such as interest in scientific phenomena and scientific activities as well as scientific literature and habits, understand the impact of science on everyday life and the society, contributions of the scientists and their sacrifices.

In addition to this, according to the Indian Education Commission 1964-66 (Kothari Commission) the objectives of teaching science at the primary level should be to develop proper understanding of the main facts, concepts, principles and processes in their physical and biological environment and to develop their power of observation.

In the lowest class, emphasis should be stressed on cleanliness and good health habits and in the next two higher classes study of personal hygiene and sanitation. In class four, they should be acquainted with the international scientific symbols for measurements and chemical elements. At the higher primary stage—classes five to seven—the emphasis is to be laid on acquisition of knowledge; and science to be taught as separate disciplines like physics, chemistry, biology, etc.

The Commission suggests adoption of disciplinary approach of science teaching rather than general science approach. At this stage, students should be trained to develop the ability to think logically and draw conclusions scientifically on the basis of evidence and observation. At the secondary level, in classes eight to ten, the Commission suggests compulsory teaching of physics, chemistry, biology and the earth sciences.

The approach should be such as to help the students develop discipline of their minds and the knowledge of science acquired should serve as a preparation for higher education in science. At the higher secondary level, there should be a distinct diversification of courses and the knowledge should serve as basis for further specialisation.

In the Commission's words, "science will not be studied on compulsory basis by all students. Those who opt for specialisation in the subject may take all the three electives from science group consisting of physics, chemistry, biology, geology and mathematics. . . . But we are not in favour of rigid grouping and a science biased course can provide for a combination of two science subjects like physics and chemistry with an arts subjects like economics.

Similarly, it is possible for an arts student to take up study of physics or biology or any other subject in the science group as an elective." The aim of teaching science can be summarised under the following categories: The students studying science should acquire knowledge of scientific facts, principles and events of nature, living and non-living, rules of health and sanitation and other kinds of knowledge of science that will help the learner to live an intelligent and efficient life in a modern society. Such knowledge must be appropriate to the age, stage and ability level of the learners.

Knowledge of science should enable the learner to understand, adjust and if necessary change his immediate environment as required. Since study without interest will not be meaningful and permanent, the teacher should therefore conduct his teaching in such a way as to stimulate interest of the learners in the subject.

The students must develop interest to pursue scientific activities within and outside the school, read scientific news and literature, organise science clubs and science exhibitions, science competitions and try to apply scientific knowledge in everyday activities. They should take interest to conserve the beauties of nature. The students studying science should be able to appreciate the contributions of science for the well-being and comfort of mankind and how the ideas and achievements of science have helped the progress of human civilization.

They should be able to realise that the tremendous progress of experimental science, the marvels and wonders of modern science, and the application of science to every field of human activity have marked the present day society as a scientific society.

The teachers in their process of teaching should lead the learners to see the impact of scicnce on society and their responsibility as a member of a modern society. The teachers should see that the desirable need for learning science is instilled in the learners' minds and that such qualities are reflected in their everyday habits. For instance, science is a pursuit of truth and its pursuit demands intellectual honesty, diligence, perseverance, tenacity, patience, concentration of mind, unbiased judgement, and objective observation.

The above said qualities also help the learners to become self-confident. Further, studies of science should develop healthy habits. While informing about health and hygiene as well as public sanitation, the teacher should emphasise their application in everyday living. The students should form the habit of taking care of personal health and community sanitation.

The knowledge of common diseases and the methods of prevention should enable learners to live a better and cleaner domestic and social life. The science classes in this respect should function as the training ground for good health habits. Through teaching of science, the students should be trained in scientific method of procedure while solving scientific problems. It involves logical steps in the process. It is, in fact, the problem-solving method which involves application of critical thinking and systematic procedure.

The scientific method means following certain steps of scientific procedure in sequential order which lead to the solution of a problem. Though there does not appear to be one strictly typical process of scientific method in general, it involves such steps as sensing a problem, defining the problem, analysis, organisation, and experimentation, collection of data, processing and calculating, formulation, of hypothesis and testing, correct interpretation, conclusion and generalisation.

The teacher should focus the learners' attention on these processes and train them to follow scientific methods to solve problems not only in the area of science but also social problems. This is an important component of the aim of science teaching. The constituents of scientific attitude may not automatically develop in the outlook of the learners.

The teacher will have to make efforts to point out these aspects in the process of teaching science and try to inculcate this outlook in the behaviour pattern of the learners.

## Scientific Attitude

Scientific attitude refers to critical observation, inquisitiveness, broadmindedness and open-mindedness, objectivity in approach, non-belief in superstition and hearsay, belief in proof, trust in correct evidence, respect for others' opinion, faith in scientific method, unprejudiced judgement, belief in cause and effect relationship. Such attitudes are warranted not only in pursuit of scientific problems but also in solving problems at home and in the society. These qualities if developed in the minds of the students and instilled in the behaviour pattern of the learners, solve many problems of individual and social living.

The scientific attitude is considered an important constituent of the aims of teaching science. Teaching of science should aim at developing in the learners various skills involved in studies of science, such as skills in drawing diagrams and sketches of specimens and apparatus, constructional skills, experimental skills for arranging and organising experiments in science as well as for handling instruments and apparatus skill of observation, and the skill for solving problems. The study of science should develop the ability to identify a problem, ability to analyse and interpret the problem, ability to improvise apparatus if necessary, ability to generalise and predict results and also the ability to locate source of information relating to the problem at hand.

The learners of science should develop abilities to express or discuss on scientific problems, the ability to organise science clubs and fairs and the ability to apply scientific principles in problems faced in life. In the scientific and technological world of today, there are innumerable avenues where science finds its application. Science, in addition to being a part of general education, prepares students for future vocations and forms the basis for specialisation in higher sciences.

The teacher should be able to select students from his group to be given special training to enable them to pursue specific discipline of science at higher level. Apart from specific studies for technical or vocational career, we need enough manpower for schools themselves. In an industrial and technological society, the use of leisure time poses a problem.

Science provides a host of technical and intellectual activities which can provide fruitful engagements for leisure. Such activities may range from production and maintenance activities to intellectual activities pursued for pleasure. While teaching facts and principles of physical and biological sciences, the pupils should be encouraged to prepare simple useful materials like soap, boot-polish, ink, etc., or articles like kaleidoscope, simple camera, simple telescope, magnetic compass etc., or engage themselves in maintenance of aquarium, herbarium, nature corner, preparation of charts, models or slides, observation of animal and bird habits, and repairing articles of everyday use and a wide range of similar other activities.

## Teaching Goals and Purposes

For carrying out any activity it is significant to know its purpose. If the purpose is clear, the activity will be carried out effectively. If one knows clearly what is to be achieved the planning will be done accordingly and the work will be carried out properly. Even if it is not carried out properly at least it can be measured or evaluated as to what extent the activity has been successful. This (evaluation) is only possible by comparing the achievement of the activity with its planned purpose, goal or objectives.

In the same way in education or in any particular field of education the identification of goals, aims and objectives is a very important activity. Here the question is what should be the aims and objectives of science education and science teaching? Before knowing the aims and objectives of science education, as a teacher we should also know that aims and objectives of teaching of any subject should be directly related to the educational philosophy and policies. Educational philosophy and policies give justification for teaching various subjects in school including science.

Aims and objectives of science education have undergone many changes because the goals of education are also changing with the changes in the society.

The objectives of science education and science teaching have been stated by various educators, education commissions and committees, and in many different terms. We talk about goals, aims and objectives, etc., sometimes they are used as synonymous terms; and sometimes they are used in different meanings. Before discussing the aims and objectives of science teaching it will be helpful at this stage to distinguish between these terms.

## Educational Aims: Classification

Once a teacher decides to teach his students, he plans several activities for a successful lesson. First he decides the topic, secondly he prepares objectives to be achieved, then he decides a relevant methodology to achieve the identified objectives. Finally he plans to evaluate student's performance.

Objectives are identified and written in various different forms. Some teachers believe that their students should "really understand" others desire that their students should know some scientific laws, facts and principles. Some desire that their students should "internalise knowledge," and still others want that their students should grasp the core or essence or comprehend. Do they all mean the same thing?

To clarify such confusions and help all teachers, administrators, professional, specialists and research workers who deal with curricular and evaluation projects, an attempt was made to build a taxonomy (or classification) of educational objectives. You will be familiar with biological taxonomies. These taxonomies are very helpful in communicational accuracy, better organisation and interrelation of various living things. Also in chemistry you will be familiar with classification of elements in periodic table.

The idea for developing a taxonomy or classification system was formed at an informal meeting of college examiners attending the 1948 American Psychological Association convention in Boston. After considerable discussion it was agreed upon that a theoretical framework which could form a common frame of reference could be used for better communication among examiners. The theoretical framework might best be obtained through a, system of classifying the goals of educational process.

For developing this taxonomy certain organisational principles were developed. It was agreed that the taxonomy should be an educational-logical-psychological classification system. Thus, *first* importance should be given to educational considerations. *Secondly* the taxonomy should be a logical classification for which every effort should be made to define terms as precisely as possible and to use them consistently. *Finally,* the taxonomy should be consistent with relevant and accepted psychological principles and theories, and value judgements about objectives and behaviours should be avoided.

Following this initial meeting at Boston in 1948, a series of meetings were held up till 1953 by the group of college and university examiners, where a threefold division of educational objectives was developed: *Cognitive affective* and *Psychomotor.* It was found that most of the objectives stated by teachers as well as found in the literature, could be placed rather easily in one of the three major domains or classifications. The three domains are separately discussed below.

According to (Bloom's) Taxonomy of Educational Objectives, the objectives can be classified into three domains:

1. Cognitive Domain (remembering);
2. Affective Domain (feeling);
3. Psychomotor Domain (doing).

***Examples***: Given below are listed some objectives of Nutrition Education in the three domains.

***Cognitive Domain (Knowledge and Intellectual Skills)***

To help the children to:

1. understand the need for adequate food and that proper nutrition is essential for physical and mental development, health and happiness.
2. become aware of the foods available in the locality and need to produce more food.
3. become aware of the ways in which food can be prepared and make attractive, tasty and nutrition in eating.

***Affective Domain (Attitude)***

To help the children to:

1. take interest in selecting, and in liking to eat different groups of food.
2. feel the necessity of producing more food and take interest in kitchen gardening.
3. take interest in finding out what kinds of food are available in the locality.
4. become convinced of avoiding waste of food at all levels.

***Psychomotor Objective (Skill)***

Children:

1. select and eat proper combination of food.
2. do not eat without washing hands, but wash hands before and after eating and develop similar hygienic food practices.
3. participate in food production at home and in school.
4. explain the kinds of food available in the locality.

## Classified Aims of Cognition

Cognitive objectives were further classified into six categories— knowledge, comprehension, application, analysis, synthesis, and evaluation.

### *Knowledge*

Knowledge is defined as the remembering of previously learned material. This may involve the recall of a wide range of material, from specific facts to complete theories, but all that is required is the bringing to mind of the appropriate information. Knowledge represents lowest level of learning outcomes in the cognitive domain.

*Examples*: Knows common terms, knows specific facts, knows methods and procedures, knows basic concepts, knows principles, theories and structure.

### *Comprehension*

Comprehension is defined as the ability to grasp the meaning of the material. This may be shown by *translating* material from one form to another (words to numbers), by *interpreting* material (explaining or summarising) and by establishing outcomes go one step beyond the simple remembering of material and represent the lowest level of understanding.

*Examples*: Understands facts and principles, interprets verbal material, interprets charts and graphs, translates verbal material to mathematical formula, estimates future consequences implied in date, justifies procedures and methods.

### *Application*

Application refers to the ability to use learned material in new and concrete situations. This may include the application of such things as rules, methods, concepts, principles; laws and theories. Learning outcomes in this area require a higher level of understanding than those under comprehension.

*Examples*: Applies concepts and principles to new situations, applies laws and theories to practical situations, solves mathematical problems, constructs charts and graphs, demonstrates correct usage of a method or procedure.

### *Analysis*

Analysis refers to the ability to breakdown material into its component parts so that its organisational structure may be understood. This may include the identification of the parts, analysis of the relationship between parts, and recognition of the organising principles involved. Learning outcomes here represent higher level of intellectual skills than comprehension and application because they require an understanding of both the content and the structural form of the material.

***Examples***: Recognises unstated assumptions, recognises logical fallacies in reasoning, distinguishes between facts and inferences, evaluates the relevancy of data, analyses the organisational structure of a work (art, music, writing, project report).

### *Synthesis*

Synthesis refers to the ability to put parts together to form a new whole. This may involve the production of a unique communication (theme or speech), a plan of operation (research proposal), or a set of abstract relations (scheme for classifying information). Learning outcomes in this area stress creative behaviour, with major emphasis on the formulation of new pattern or structures.

***Examples***: Writes a well organised theme, gives a well organised speech, writes a creative short story (or poem or music), proposes a plan for experiment, integrates learning from different areas into a plan for solving a problem, formulates a new scheme for classifying objects (or events or ideas).

### *Evaluation*

Evaluation is concerned with the ability to judge the value of material (statement, novel, poem, research report) for a given purpose. The judgements are to be based on definite criteria. These may be internal criteria (organisation) or external criteria (relevance to the purpose) and the student may determine the criteria. Learning outcomes in this area are highest in the cognitive hierarchy, because they contain elements of all of the other categories, plus conscious value judgements, based on clearly defined criteria.

***Examples.*** Judges the logical consistency of written material, judges the adequacy with which conclusions are supported by data, judges the value of work (art, music, writing) by use of internal criteria, judges the value of work (art, music, writing) by use of external standards of excellence.

## Influenced Aims Classified

Affective domain objectives deal with interest, attitudes, values, appreciation and adjustment. These objectives describe behaviours ranging from students merely being aware that a given phenomenon exists to behaviours where they are increasingly willing to attend, respond to phenomenon with a positive feeling, attach values with it and at some point conceptualise their behaviours and feelings.

Finally they organise these conceptualisations into a structure. This structure grows in complexity as it becomes his life outlook.

This shows a process by which a given phenomenon or value passed from a level of bare awareness to a position of some power to guide or control the behaviour of a person.

Affective domain objectives were further classified into five categories intended to be hierarchical in order.

### *Receiving (Attending)*

Receiving is to orient the learner to learn. This is the first step that he is willing to learn what is being given to him.

The category of receiving has three different levels, the lowest level is the awareness or being conscious of something. The next point is that he is willing to receive and then he is attentive when something is given to him.

***Examples:***

1. The students are aware of scientific activities in the school.
2. They are willing to take part in scientific activities.
3. They attend activities arranged by science club.

### Responding

Responding level comes after the learner has given his attention. Responding includes behaviours like compliance and willingness to respond and getting satisfaction by responding.

***Examples:***

1. Students respond to the questions asked by the teacher.
2. Students raise hands to answer questions every time a question is asked by the teacher.
3. Students feel happy after answering the questions correctly.
4. Students take pleasure in explaining scientific phenomenon to others.

### Valuing

The abstract concept is an important element of behaviour. At this level individual is not motivated by the desire to comply or obey but he is motivated because of his own valuing or assessment and commitment to the underlying value. The value system has slowly accepted and has come to be used by the student at his own criteria of worth.

***Examples***: Students have faith in the power of reasoning and methods of experimenting and discussions.

### Organisation

When the learner develops certain values, he encounters situations for which more than one value is relevant. In such cases he is able to organise the values into a system.

***Examples***: Students develop a plan for doing some scientific activities and school work at home.

### Characterisation by a Value or Value Complex

The individual acts constantly in accordance with the values he has developed.

***Examples***: Students develop a code of behaviour based on scientific attitudes.

## Influenced and Cognitive Domains

It is really hard to compartmentalise human behaviours clearly in terms of cognition and affect. No teacher or curriculum developer really intends one entirely without the other. There are many researches which demonstrate that cognition and affect can never be separated.

It is also noticed that the interest arises from increased information about something. According to Burner, some workers have felt that it is the process of problem solving and discovery in learning that brings about increased motivation for the subject and therefore develops interests and attitudes. Their view is that it is not much what is learned, but how it is learned which will determine the affective objectives that will be attained at the same time as the cognitive objectives.

Thus, you learned about the aims and objectives of science teaching. As a science teacher you should distinguish between "aims" and "objectives" and "general and specific objectives" of science teaching. You will be teaching science at various levels — middle, secondary or senior secondary. Now you can identify objectives of science teaching at various levels.

Now you know that the objectives of science teaching can be placed into three domains — cognitive, affective and psychomotor. As a teacher, for your science lessons, you can take objectives from each domain. Identify only those objectives which you feel you can achieve. They should be compatible to the cognitive level of students and the existing classroom conditions.

If objectives are written in behavioural terms, teachers know what exactly they are to teach, students know what exactly they are to learn, textbook writers know what exactly they are to write and paper setters know what exactly they are to ask in examination.

## The Contrast

Richard Whitfield of Faculty of Education, Cambridge has described these in the form of a summary for the isolation of teaching objectives. This summarised form will give us a clear idea of the various terms the way they are quite often used.

## The Objectives

The first two steps are the goals or aims; they serve to suggest general directions in which our science courses may be steered.

Justification for science in an educational setting Suggests

↓

(1) Aims for science teaching
Translation and selection gives

↓

(2) Objectives of particular courses
By selecting and relating these to specific content and teaching methods, we obtain

↓

(3) Objectives for particular lessons
Some of these at least are amenable to behavioural specification, i.e., identifiable in terms of observable qualities of pupils.

1. They are general global targets (general objectives) with which we are concerned.
2. They form the basis of the general place of science in the curriculum for given groups of pupils.

*Examples*

1. To develop interest of pupils in science.
2. To develop enquiring minds.
3. To help pupils to see science in relation to the rest of culture.
4. To prepare some pupils for carriers in science.
5. To help pupils to solve problems

**Objectives (Step 3)**

1. They are the more precise outcomes (specific objectives) for a particular course of teaching or even a single lesson.

2. They are derived from the aims and represent their translation into terms which are tangle in planning a course or series of lessons.

* When objectives written in this form, they are called behavioural objectives.

***Examples:***

1. To develop the ability to formulate hypotheses.
2. To foster the ability to estimate experimental errors.
3. To develop skill in wiring simple circuits.
4. To increase pupils' ability to work as a term in sharing out a complex task.
5. To help pupils describe the structure of human heart.

An objective is an intent communicated by a statement describing a proposed change in a learner — a statement of what the learner is to be like when he has successfully completed a learning experience. It is a description of pattern of behaviour (performance) we want the learner to be able to demonstrate.

In the given examples of objectives some objectives can be achieved by one particular lesson and others can be achieved by many. Such objectives which are very specific in nature and can be achieved by a particular lesson are called *specific objectives*, others which will be achieved by teaching many or even the whole course are called *general objectives.*

In this way, Richard Whitfield explains that the clarification of aims and objectives is what distinguishes the 'professional' teacher from the technician carrying out someone else's orders. The discussion also shows how general aims and justifications for teaching science can be translated into more detailed statements of objectives which affect the choice of curriculum content and teaching methods.

## State-wise Programmes

There is hardly any need today to justify the place of science in a scheme of general education for school children.

Science is all pervasive. Modern societies exist on the basis of science; science is intimately related to the means of production and means of communication, including transport. Even economics and politics have to depend on scientific factors such as productivity from the land or from industry, the power of modern weapons or the speed of transportation of ground and air forces. In the present situation, therefore, anyone, in any walk of life, must be aware of a certain quantum of science and technology.

The scientific method extends far beyond science. All disciplines are becoming scientific. The method of observation, of making symbolic, graphical or linguistic models, of designing experiments, of applying reason as well as imagination to draw conclusions from data to formulate theories/ the method of keeping an objective view while theories are tested is a method which pervades every discipline. The facts of today may not be the facts of tomorrow, and theories may undergo change, but there can be no going away from the scientific method.

Science should help in reducing obscurantism and all sorts of prejudices based on sex, caste, religion, language or region. By emphasising a rational approach science should help the development of a democratic, secular and socialist state.

The general objectives of science education have a limited utility unless they are spelt out in terms of stage-wise objectives for the guidance of educational planners, administrators, supervisors and teachers. According to Kothari Commission Report (1964-66), NCERT has laid down the following objectives as national objectives of science education in its publication, "The Curriculum for the Ten-Year School — A Framework," for various levels.

### *Primary (I-V)*

This is a very crucial stage in the life of a child. The child's spontaneity, curiosity creativity and activity in general should not be restricted by a rigid and unattractive method of teaching and environment for learning. The curriculum should take into consideration the social, intellectual, emotional and physical maturity of the child as well as the socio-economic needs of the community.

It will be helpful to identify realistically the basic minimum to be achieved in respect of each and every child and leave enough scope for individual schools to go as far beyond this basic minimum as their circumstances permit. There should be enough scope for flexibility and local adjustments. As far a number of children the primary stage may be terminal, it should be necessary to provide an education to them which prepares them for life and for self learning.

In the primary classes, the sciences should be taught as environmental studies; in Classes I and II as a composite course including both the natural and the social environment, and later on, in classes III, IV and V, as two subjects—environmental studies I (natural science) and environmental studies II (social science). One needs not lay down how much of this should be covered in a particular class. The purpose should be not to stuff the minds of children with facts and information, but to sharpen their senses, to enable them observe their environment and to enrich their experience.

Since the environment and the experiences of the children outside the school vary from place to place, the activities provided in the school should also vary so that edifice of knowledge is built not on abstract concepts but also on the solid foundations of experience drawn from the environment of the child. In such a programme a flexible syllabus should be drawn up to cater the needs of all sorts of schools. There is always a criticism that our science textbooks have generally an urban bias.

The textbook should not be given so much importance in the learning of science. What is more important at this stage is to provide teachers with guides and prepare them to develop their own instructional material. Reading materials for children should be prepared very thoughtfully so as to motivate them for reading.

The general objectives of science education at primary stage may be stated as follows:

1. The child should learn the method of enquiry in science and should begin to appreciate science and technology in everyday life.
2. The child should develop habits of cleanliness and healthful living and an understanding of the proper sanitation and hygiene of his neighbourhood.
3. The child should learn to cooperate with others and appreciate the utility of working together. Other desirable qualities of character and personality, such as initiative, leadership, honesty should be developed.
4. The child should be able to express himself freely in creative activities and should acquire habits of self-learning.

*Middle (VI-VIII)*

At this stage the children become adolescent and this period can become difficult for many children. Problems of adjustment in the family, the school and the society begin to appear. The child becomes a boy or a girl with greater intellectual, emotional, social and physical maturity than the primary school child. Social demands and responsibilities begin to appear. For many boys and girls, this stage is terminal after which they enter life and work. Therefore, they should be prepared adequately to face life and develop capacities and attitudes for productive work in which they have to participate.

In the sciences, at this stage, physical and life sciences should be introduced. At the same time/ environmental education, nutrition, health and population education should receive adequate attention so that science is related meaningfully to life.

This is the age when work experience should emphasise agricultural and technological processes and tools to help the integration of science, mathematics and technology with production and with the life of the community.

In middle classes (VI-VIII) a number of States and Union Territories have already gone ahead with the introduction of physics, chemistry and biology as disciplines. Some other States and Union Territories could develop integrated science courses upto class VIII; some could also try with groups — physical sciences (including physics and chemistry), and life sciences (including botany, zoology and human physiology).

The unit approach, which does not violate the disciplines, and actually brings them closer and provides a more logical and economical procedure, has a great potential and should be adopted. In these classes laws and theories should be gradually introduced, taking care that they are not introduced as dogmas. Pupils should be made to understand that many models are available and one of them is the one they are learning. Open-mindedness and scepticism should be encouraged.

The general objectives of science education at middle stage may be stated as follows:

1. The child should be able to apply the knowledge of science in everyday life.
2. The child should be able to investigate new knowledge in the field of science.
3. The child should develop scientific attitude.
4. The child should learn how to learn a part of scientific knowledge on his own.
5. The child should be able to solve problems around him.
6. To make the child creative.
7. To train the child in science processes.

***Secondary (IX-X)***

The secondary stage covers the two classes, IX and X. These two classes complete the ten years of general education. After this, there are three possible courses open to students:

1. They can enter the working force,
2. They can take up vocational courses, and
3. They can take up higher level academic courses of study to prepare for entrance to the first degree class in the college or university.

The IX and X classes may be terminal for a large majority of students who are to enter the working force. Therefore, it is necessary to bring their education upto a standard which will give them the competence to enter life. The Education Commission (1964-66) had thought of a minimum national standard of attainment so that those areas of the country and those sections of the society which are backward could come up and compete fairly with the rest of the people.

Thus, for maintaining a continuity of the objectives of education from the previous years, it is necessary to pay attention to the academic subjects as well as to the knowledge and skill required for doing socially productive work. These two years are crucial from the point of view of the development of the personality.

While from the onset of puberty in the middle stage there are problems of adjustment which the young child has to cope with. It is at this stage that these become acute and the additional preparation for a transition from the life of a school student to that of a productive person in society has to be undertaken.

Therefore, it is necessary to give the child some psychological insight into his problems and the knowledge that may help him to understand his own behaviour as well as that of others around him.

The general objectives of science education at this stage may be stated as follows:

1. The student should develop the competence to apply his knowledge to the solution of the problems around him.
2. He should have an understanding of the technological processes in agriculture and industry in use in his surroundings.
3. He should be able to contribute meaningfully to environmental conservation, the reduction of pollution, the development of proper nutrition and health and hygiene in the community.
4. He should be able to help in the development of proper habits and attitudes in the child care and in the improvement of the home.
5. He should acquire the knowledge and skills required for entry into an area of work.
6. He should learn one or two useful trades.
7. It is equally important to give him enough knowledge of the materials, tools, techniques and processes of a job family so that he can enter life with some confidence.
8. Through curricular and co-curricular activities (science club, science fair, science debate, science essay, science symposium, etc.) he should develop desirable scientific attitude and values like cooperation, team spirit, fellow feeling, leadership, courage, truthfulness, honesty and sincerity.

***Senior Secondary (XI-XII)***

You will be teaching physics, chemistry or biology or just general science or integrated science. Here is a list of objectives of science teaching as identified by a group of science teachers (in a workshop at SCERT Delhi) teaching science (Physics, Chemistry or Biology) to senior secondary classes (XI-XII). Some of them may also fit in the lower classes (middle or secondary). You may or may not agree with all the objectives. You might like to add some more adjectives to this list.

Thus, this list of objectives of science teaching may be a guideline for you to identify some objectives which you will like to achieve at the end of your lesson, unit, course or programme. These objectives are listed below:

1. To give students up-to-date knowledge in science.
2. To develop their curiosity to learn science.
3. To train them in scientific method of learning.
4. To orient them to apply science and technology in agriculture, medicine, research and industry.
5. To develop in them the competency to apply scientific knowledge to the solution of the problems around them.
6. To enable them to contribute meaningfully to the conservation of environment, the reduction of pollution, the development of nutrition, health and hygiene conditions in the community.
7. To enable them to help in the development of proper habits and attitudes in child care and in the improvement of the home.
8. To enable them to reduce all sorts of prejudices based on sex, caste, religion, language or region.
9. To sharpen their senses to enable them to observe their environment and to enrich their experience, rather than to stuff their minds with facts and information.
10. To make them aware of these aspects of science that are environment based and life centred.
11. To train them in the use and maintenance of science equipment.
12. To make them creative.
13. To develop in them the mental faculty of open-mindedness, making them understand that there are many models of experimentation that are possible.

14. To prepare them for change and create in them confidence for facing problems of the future.
15. To enable them to formulate and test hypotheses.
16. To enable them to investigate the scientific phenomenon rather than just verifying known conclusions.
17. To enable them to investigate the scientific laws, principles and facts, etc., which they can do by laboratory experiments.
18. To enable them to believe and remember scientific laws, principles, and facts, etc., discovered by others and in the textbook.
19. To train them in science processes in order to discover some of the science content.
20. To involve them in learning processes as far as possible instead of treating them as passive listeners.
21. To build proper base for professional and higher science courses.

# 5

# Organised Teaching

Percentage of our science teachers (Physics, Chemistry and Biology) have good background in their respective subjects, and they take interest in teaching and trying out new science programmes. Even in the schools which have enough science equipment and the science labs are well equipped, the lecture-demonstration method is the one most commonly used, as our science teachers are heavily loaded and they get a very little or sometime even no time for planning their lessons during school hours.

Our schools have usually 8 periods of 30-40 minutes duration. Regular teachers also work as substitute teachers (where in America if a teacher is on leave, a substitute teacher is invited from outside to teach his classes) during their planning periods besides their teaching 6-7 (out of 8) periods daily, six days a week (where in Western countries schools work five days a week), and Sunday is the only holiday.

Our textbooks are written in a traditional way. If our science teachers could have well sequenced programmed materials, it would be a great help to them, as mentioned above, they are heavily loaded. Also if the programmed materials provided activities of 20-25 minutes duration, students would have more opportunity to do the experiments themselves even in smaller periods. So well sequenced programmed materials would be more useful for teachers to teach and for students to learn than the materials now available.

**Origin and Development**

In developing this unit the System Approach was used. The System Approach has the following steps:

1. Identify problems and state terminal objectives.
2. Conduct task analysis.
3. Describe entry behaviours of students.
4. State sub-objectives in behavioural terms.
5. Develop evaluation instruments.
6. Determine instructional sequence.
7. Select appropriate media and instructional procedures.
8. Develop instructional materials.
9. Conduct formative and summative evaluation.

**Hindrance (Terminal Objective)**

Given the atomic weight and atomic number of an element the students will find out the number of protons and neutrons in the nucleus and the number of electrons in the various shells of the atom.

**Activity Interpretation**

Hierarchical methodology was chosen for the programme structure of an atom. There is an ordered relationship between sub-skills which ultimately help to learn the terminal objective. The simple things are given first that makes the basis for more complex concepts.

**Entry Attitudes**

The assumed entry behaviours for this programme were the following:

1. Knowledge of some common elements such as oxygen, hydrogen, nitrogen, carbon, sulphur, sodium, chlorine, etc.

2. Simple addition, subtraction, multiplication and division.
3. Knowledge of positive and negative charges.
4. Basic skills of reading, writing, generalising, and following directions.

**Purpose of Attitude**

Behavioural objectives were prepared in such a way that they were consistent with each task in the task analysis. All the behavioural objectives contain an observable response, important condition under which the behaviour will be evaluated, and the criterion for satisfactory performance.

**Weighing Instruments**

The test was made to test all the eleven behavioural objectives. It contained 16 test items. There were one or two test items to test each behavioural objective. All the test items are multiple choice tests and seems to be appropriate to test the task required in the behavioural objectives. The language used in writing test items was kept easy enough to be understood by the students. This test was used as a Pre-Test, in which the items are sequenced in the same manner as they would appear in the programme (in the Sequence of Tasks and Behavioural Objectives). The same test can also be used as Post-Test (though a separate Post-Test is preferred), whose items should be scrambled, so that they are not in the same order as in the programme. In the multiple choice questions the order of correct responses was random.

**Instructional Series**

The frames were constructed in the same sequence as shown in the task analysis. The frames were sequenced in such a way that by completing all the frames the terminal objective might be achieved by the students.

**Suitable Means**

The strategy of linear programming was chosen. The self-learning unit was an instructional programme, having enough practice frames.

## Programme Evaluation

### *Formative Evaluation*

After writing the programme and the tests (pre-test and post-test) one IX class student was selected for one-on-one evaluation. This student was of an average ability. Two copies of the programme were prepared, one for the student and other for the programmer. When the student was using the programme, the programmer noted down his reactions and suggestions, which seemed to be important in revising the programme. The following feedback was received from one-to-one evaluation.

1. There was smooth flow or material in the programme, that is, the arrangement of the frames was logical.
2. At some places difficult language was used.
3. The programme needed more practice frames.
4. Frame No. 18 was not well explained, cue and prompt were also needed in this frame.

The programme was revised according to the feedback received from one-on-one evaluation. Then the programme was tried out on a small group consisting of five IX class students. Each student was given pre-test, the programme and the post-test. Before starting the programme the group was told about the programme and the importance of the criticism.

They were also asked to write any suggestions if they have. They all took 40-55 minutes to complete the programme. Finally they were interviewed by the programmer and were asked to give their suggestions. The programme was further revised according to the feedback received from small group evaluation.

### *Simmative Evaluation*

The revised programme was given to the target population, (150 IX class students) after giving the pre-test. There were few students who asked questions when going through the programme, but when they were asked to read the matter again they seemed to understand it. All of them were able to complete the programme in time (about 2 periods of 30 minutes duration). Post-test was given the next day.

The pre-test and post-test gain shows that 72 per cent of the students achieved 80 per cent or more objectives. The majority of the students used in the summative evaluation programme showed enthusiasm in working with the programme. Many of them said that they liked the new system of instruction and wanted to have more programmes like this. The vast pre-test to post-test gains exhibited by most of the students generated a great deal of excitement and confidence. This fact alone made the whole project a worthwhile experience for the students.

## Some Facts about the Unit

### *Constructed Response Frame*

As the word Constructed implies, no choice are presented to the student in a Constructed Response Frame. He does not select one response from many, as in the Discrimination Frame. Instead, the student constructs his own response each time. That is, he supplies the answer from his own knowledge.

The response the student constructs can take many frames. He may be asked to write or supply a word or statement, draw a diagram, or perform any other type of over action requiring a response from within his own repertory.

### *Examples*

Frames No. 1, 2, 8, 11, 12, 13, 14, 16, 17, 18, 19, 20, 21, 22 and 23. The Constructed Response Frame is basically a two-part structure, the set frame and atleast one *practice frame.* It may be desirable to have several practice frames with each set frame.

### *Set Frame*

Whenever the response asked for is found in the data portion of the frame, it is known as a set frame. The student may never have seen the desired response prior to reaching this frame, but he is able to supply this response simply by deducing it from the data within the frame itself.

***Examples:*** Frames No. 1, 2, 11, 12, 13, 14, 16, 18 and 21.

### *Practice Frame*

The set frame is followed by a Practice Frame. The practice frame gives the student a chance to practise what he has learned in the set frame.

***Examples***: Frames No. 8, 19, 20, 22 and 23.

***Discrimination Frame.*** As the name implies, this construction technique is used to teach student to make discrimination. A student who has been taught to make fine discriminations through the use of Discrimination Frame sequence will find that he can go beyond the programme material and approach the subject matter from any direction. For example, if taught a definition by Discrimination Frame Sequence, he will be able to define it when given the term; if given the definition, he will be able to name the item; if physically possible, he will be able to illustrate it.

***Examples:*** Frames No. 3, 4, 5, 6, 7, 9 and 15.

### *Baboon Frame*

In this frame the student is asked to make a choice from among four answers: Choice A, Choice B, both A and B, or neither A nor B.

***Examples:*** Frame No. 10.

## More About Baboon Frame

Basically, the BABOON Frame Sequence consists of three frames similar in purpose to those of the Constructed Response Frame Sequence. The first frame is a set frame; it contains enough information to enable the student to come up with the correct response when asked to respond. This frame is followed by a practice frame; here the student is asked to demonstrate, with a little prompting. In the final frame, the minimum amount of stimulus is presented to the student and a maximum response is called for.

*Examples:*

1. A Trapezium is a figure with four sides, two of which are parallel. Place a Tick mark (✓) before the correct statement below tick only one Answer:
   (a) Figure A is a trapezium.
   (b) Figure Â is a trapezium.
   (c) Both Figure A and Figure Â are trapeziums.
   (d) Neither Figure A nor Figure Â is a trapezium.
2. Place a tick mark (✓) before the correct statement below. tick only one answer:
   (a) Figure A is a trapezium.
   (b) Figure B is a trapezium.
   (c) Figure A and Figure B are trapeziums.
   (d) Neither Figure A nor Figure B is a trapezium.
3. Define "Trapezium" and draw two figures. A trapezium is a figure with four sides, two of which are parallel (or words to this effect). The example figures that you have drawn should match this definition.

## Typology of Programmed Units

### *Linear Programme*

In linear programmes, all of the students are normally required to take all of the frame.

*Example:*

The self-learning unit, "Structure of an atom" is a linear programme.

Linear Programme strategy was used in writing this Programme due to the following reasons:

1. The topic was new for most of the students for whom it was developed, and they were needed to be provided with a lot of practice.

2. The topic of the programme was of such nature that the subject matter involved ascending order of complex skills to be learned/ therefore it was felt that all the students should be required to take all frames.
3. The type of learning task necessitated mostly constructed response answers.

### *Branching Programme*

The word branching suggests any deviation from the straight line, A Branching Programme allows for greater differences in student abilities.

### *The Branching Frame Sequence Technique*

This technique presents the student with remedial information/ if necessary and permits him to take steps that are as large as his capabilities allow. A particularly adept student may go through a programme in a minimum number of steps, whereas his less able cohort may require twice that number to learn, the same amount of material. The Branching Programme offers the student alternate paths from which to choose, and the path he takes depends upon the response he makes in each frame.

### *Improved PI Units for Exercise*

Hence now you know how to develop a programmed instructional unit. Identify some topics in your field, on which you feel good programmes may be developed. Develop, evaluate and revise programmed instructional units on these topics, and share these self-learning units with your fellow teachers.

## General Aspects

Let us look at some of the definitions of programmed instruction:

1. Cronbach describes a programme of instruction as follows: "A programme is a pre-arranged sequence of explanations and questions. A programme whether for a brief unit or

for an entire course, is a carefully planned progression of ideas, beginning with elementary notions and working upto relatively complex theories or applications.

2. According to B.F. Skinner a programmed instruction is simply a matter of breaking the material to be learned into easy steps, arranging steps in logical order with no gaps, making sure the student understands one step before moving to another and then incidently, making sure that he is successful.
3. According to Espich and Williams a programmed instruction may be defined as a planned sequence of experiences, leading to the proficiency, in terms of stimulus response relationships. By this definition a programme is an educational device that causes a student to progress through a series of experiences, which lead to the students proficiency. The experience here is student's own experience in the learning process, not just the teacher telling. Planned sequence determines what experiences and in what order should occur. What is the student supposed to be able to do after completing the programme? How well? How quickly? All such questions implied in leading to proficiency in terms of stimulus-response relationships refer to the basic behavioural science concepts on which programmed instruction is based, and which are taken into consideration when a programme is written.

## Psychology Associated with Behaviour

No one knows for certain how or why programmed instruction works, but it is generally agreed that basic behavioural psychology is somehow involved. At least the originators of programmed instruction attributed its success to some basic tenets of behavioural psychology.

The results of the change in behaviour called learning are observable or measurable. All behavioural changes of students as a result of learning may be of three types: (1) Psychomotor;

(2) Cognitive; and (3) Affective. (1) and (2) are comparatively easier to measure than (3).

In any teaching situation for effective learning the best method is:

1. Present the stimulus to the student.
2. Help the student to make the desired response to the stimulus by giving him clues, by leading him towards it, or by telling him the response itself.
3. When die student makes the desired response to the stimulus, immediately reinforces that response.

Programmed instruction takes advantage of the basic human drive for success. The programmer guides the student towards making the correct response. He then shows or tells the student that he has given the desired response — that he has been successful. Each time he makes the correct response, he is positively reinforced by being told that he is correct, his drive for success is satisfied. Each time his drive for success is satisfied, the probability increases that he will make the correct response to the given stimulus in future situations.

## Significances of Programmed Guidance

The programmed instruction does have a number of advantages over conventional methods. It allows for individual rates of learning and it gives immediate reinforcement of correct responses. According to Burner the technically most interesting features of automatic devices are that they can take some of the load of teaching of the teacher's shoulders.

For this time the teacher can be used by his students who need him and whenever they need him. According to James this type of teaching is very frustrating and tiring but extremely rewarding. When science teachers have worked as long at individualised instruction as they have to make grouped instruction workable, the rewards may be proportionally increased.

## Combining Practicals with Theory

In teaching science, a continuing area of concern to educators has been the problem of integrating laboratory experimentation with instruction in scientific theory. The need to individualise course content for students is also recognised. Programmed instruction may offer one solution to both of these difficulties. If the science curriculum can be programmed so that individual rates could be dealt with more effectively, and if laboratory material can be developed which enable students to conduct experiments effectively on an independent basis, laboratory experimentation may be integrated more satisfactorily into the typical science course.

A summary of the subject area in which programmed instruction has been used reveals that these areas which require laboratory activities are rare. Cowan indicated that there were no auto-instructional materials available in physics that provided students with laboratory experiences. Cowan and Siddiqi developed and used such materials in their research studies.

## Additional Programming

Hundreds of good science textbooks are available in the market. They are not designed to teach but to convey information to the student. In a programme a programmer determines for a student, what he should and what he should not assimilate. In a programme, the student is guided along a path and given those experiences that will cause him to learn those things. No such guidance is given with a textbook. Adjunct programming can be a link between programmed instruction and a good textbook. It combines some of the progressive features of programmed instruction with the comprehen-siveness of textbooks. The goal of adjunct programming is to enable the student to learn as effectively as possible from a good textbook.

An Adjunct Programme may be one of two types (1) The text itself is kept intact and the programme is supplied as a separate unit, or (2) sections of the textbook are extracted vibration and used in the programme as the basic information. The most popular procedure as used by Cowan and Siddiqi is to leave the textbook intact. Their programmed materials which they developed for the PSSC Physics Text, provide a very good example of adjunct programming.

They used these materials in their research studies. Students using the auto instructional materials studied from the same text, worked the same problems, viewed the same films, and performed the same experiments as regular physics students. Cowan did not use a qualified physics teacher in the classroom to direct their studies.

The use of auto instructional materials prepared for this study evidently provided a good physics programme, as students using these materials were able to demonstrate as well in achievement as those not using these materials but taught by a qualified PSSC physics teacher.

In Siddiqis study students of experimental group used the auto-instructional materials under the guidance of their physics teacher, while, the students of the control group did not use such materials, they were taught by their teacher in a conventional way. The experimental group demonstrated statistically higher mean level of achievement in PSSC physics.

### *Students' Opinion about Programmed Instruction (PI)*

Siddiqi developed an Adjunct Programme for PSSC physics. It was student self-directed study guide in PSSC physics. Here are the comments (unedited) of some of the students who used it.

### *Strengths*

1. "Students can do whenever they want to do. If we do not understand anything now, we can do it some other time."
2. "When you learn on your own, you are more apt to remember it."

3. "Student learns more if he goes on his own rate. In conventional classes superior students get frustrated. They want to go ahead. Slow students don't understand. Teacher stops the whole class. With these materials superior students also have an opportunity to go ahead. Every student is able to go on his own rate, not waiting for anybody, not trying to catch up with anybody."
4. "Teacher gets more time, when he can go and help other students. He goes to the student who has problems, and not to the student who does not have any problem; so he does not waste his time. Teacher's time can be put to better use. We do not need the teacher all the time. We only use him when we need him. Teacher can enrich the student more when he talks individually."
5. "PSSC text was easy to understand with these materials. Text was too complicated, and the guide made it simplified."
6. "Guide has more hints to do labs than does the lab manual."
7. "Self-tests were good. When I missed something, I went over it and it helped me very much to learn physics."

***Weaknesses***

1. "These materials were unable to motivate those who were not self-motivated to use these materials."
2. "This approach did not work for lazy students. Some students need a push and without a teacher always available to keep them working they have a tendency not to study. It's hard for some people to pace. It requires more discipline to work on your own."
3. "This approach is good for those who want to learn by this approach. If the student is the type who cannot go on his own, this approach is not good; he should (probably) be taught by the conventional method."

4. "Sometimes a student is isolated by this approach — text guide, problems, lab, text, guide; that's all he has to work with."
5. "Physics is such a hard subject that you have to have more teacher help."
6. "I did not like it too much, because I am lazy by nature."

***Teachers' View about Programmed Instruction (PI)***

Teachers who taught the students using programmed instruction materials also pointed out some strengths and some weaknesses of misapproach of teaching.

***Strengths***

1. "These materials are excellent for those who have difficulty with PSSC physics."
2. "These materials enabled some students to achieve what they were unable to achieve without them."
3. "These materials developed in students an attitude of doing independent work, which they will need when they will go to College and University."

***Weaknesses***

1. "Only those students who feel responsibility learn more by this approach."
2. "These materials are not good for slow learners."
3. "Some students go behind and behind not because they were not capable to move but because they were lazy. Some students can read but they don't want to read. There is a problem of keeping the student motivated."
4. "This approach is more time consuming. Some students took more time to finish the same material."
5. "When working on their own only half of the students use their time effectively while others use their time doing something else."

***Assuming Remarks of Teachers and Students for (PI)***

The conclusion of a group of students regarding the autoinstructional material was: "With autoinstructional materials, the text, and a teacher to explain, the self-study would be most practical and best for students."

In general most of the students and teachers liked the programmed instructional materials and this approach of instruction. They pointed out individualisation, independence, self-pacing, self-evaluation, and better use of teacher's time as some of the major strengths of this type of instruction. More than half of the students and teachers liked individualised instruction. They reported that using this approach for instruction, the teacher got more time to interact with individual students. They pointed out that the PSSC Physics Students Guide made PSSC Physics easier for the students. They also pointed out that this approach of instruction was best suited for superior students. More than half of the students liked the idea of moving according to their own pace, but the teachers did not like this idea very much, as it was hard for them to cope with their students, when different students were doing different things in the same period. Most of the students and all the teachers pointed out that lack of self-motivation was the main weakness of this type of instruction.

## Flourishing a Programmed Guidance Unit

Some of you as a science teacher will like to use programmed instruction materials in your science classes. Usually programmes are not available in the market. Though developing programmed instruction materials is a time consuming job, yet some of you will like to develop some programmed instruction units. It will be a good project for your instruction.

Before developing programmed instruction units, it will be nice if you learn how to develop them. One such programmed unit "Structure of an Atom" is given here as a Sample. You go through it, and after that we will discuss, how it is developed.

***Example:***

**Structure of an Atom**
**(A Programme Unit)**

***Instructions to Students:***

This is not a test. This is a Programme designed for you to learn the Structure of Atom.

In this Programme you will find numbered paragraphs.

These paragraphs are called Frames. Read each Frame carefully and answer the questions given after each Frame.

Check your answers from the Answer key given at the end of the programme.

***Frame 1***: Matter is made up of very small particles called atoms. The smallest particle of matter is........

If answer is correct go to Frame 2, otherwise Repeat Frame 1.

***Frame 2***: Atom is made of various particles called electrons, protons and neutrons. Electrons are negatively charged particles. Protons are positively charged particles. Neutrons have no charge, that is, they are neutral.

Complete the following sentences:

(a) The positively charged particle of an atom is.....

(b) An electron has ...... charge.

(c) A neutral atomic particle is called.......

If answers are correct go to Frame 3 otherwise Repeat Frame 2.

***Frame 3***: Check the statement below that is true. A proton is:

(a) positively charged;

(b) negatively charged;

(c) neutral;

(d) sometimes positive, sometimes negative.

If answer is correct go to Frame 4, otherwise Repeat Frames 2 and 3.

*Frame 4*: Check the statement below that is true. A neutron is:

(a) positively charged.

(b) negatively charged.

(c) neutral.

(d) sometimes positive, sometimes negative.

If answer is correct go to Frame 5, otherwise Repeat Frames 2 and 4.

*Frame 5*: Check the statement below that is true. An electron is:

(a) positively charged.

(b) negatively charged.

(c) neutral.

(d) sometimes positive, sometimes negative.

If answer is correct go to Frame 6, otherwise Repeat Frames 2 and 5.

*Frame 6*: The weight of a proton is approximately equal to the weight of a neutron, and the weight of an electron is so small compared to the weight of a proton or a neutron that it can be neglected.

Check the statement below that is true.

The weight of a proton is approximately equal to the weight of:

(a) electron.

(b) neutron.

(c) each of them.

(d) none of them.

If answer is correct go to Frame 7, otherwise Repeat Frame 6.

*Frame 7*: The charge of an electron is equal and opposite to the charge of a proton. If the charge of an electron is -4 units, then the charge of a proton should be:

(a) –4 units

(b) +4 units

(c) more than 4 units

(d) less than 4 units

If answer is correct go to Frame 8, otherwise Repeat Frame 7.

***Frame 8*:** If the charge of a proton is $+x$ units, the charge of an electron should be.......

If answer is correct go to Frame 9, otherwise Repeat Frames 7 and 8.

***Frame 9:*** An atom can be assumed to be spherical in shape. At its centre there is a very small space (compared to the atom as a whole) containing protons and neutrons. This space is called nucleus.

Check the statement below that is true. The nucleus of an atom contains:

(a) protons only.

(b) neutrons only.

(c) protons and neutrons both.

(d) none of them.

If answer is correct go to Frame 10, otherwise Repeat Frame 9.

***Frame 10*:** Around the nucleus some hollow spheres of different sizes can be assumed. The space between any two of such hollow spheres is called a shell

Check the statement that is true.

(a) A is a shell.

(b) A is a shell.

(c) Both A and Â are shells.

(d) None of them are shells.

If answer is correct go to Frame 11, otherwise Repeat Frame 10.

***Frame 11*:** As the number of electrons is equal to the number of protons and the charge of an electron is equal and opposite to the charge of a proton the atom as a whole is neutral.

***Example***—Sodium atom has 11 electrons and 11 protons. If charge of an electron is $-x$ units, the charge of a proton will be $+x$ units.

Charge of 11 electrons = $-11x$ units.

Charge of 11 protons = $+11x$ units.

Therefore, total charge of sodium atom is equal to $-11x$ plus $+11x = 0$.

Complete the following statements:

(a) A neutral atom has 40 electrons, then it should have.......... protons.

(b) If an atom has 10 electrons but only 9 protons, then the atom...... neutral (will be, will not be).

If answers are correct go to Frame 12, otherwise Repeat Frame 11.

***Frame 12***: Nucleus has protons and neutrons. Around the nucleus there exist electrons in various spherical shells.

Complete the following statements:

(a) Electrons exist in the.........

(b) Protons exist in the..........

(c) Neutrons exist in the..........

If answer is correct go to Frame 13, otherwise Repeat Frame 12.

***Frame 13***: In a particular shell the electrons are assumed to be moving in various circular orbits. The total number of electrons in all the shells of an atom is equal to the number of protons in its nucleus. This number, i.e., the number of protons or electrons in an atom, is called atomic number.

(a) Sodium has 11 protons in the nucleus. The total number of electrons in all the shells of sodium will be........

(b) The atomic number of sodium is.......

If answers are correct go to Frame 14, otherwise Repeat Frame 13.

***Frame 14***: The atomic number of oxygen is 8. What will be the number of protons in the nucleus, and the number of electrons in the shells?

(a) Number of protons is......

(b) Number of electrons is.......

If answers are correct go to Frame 15, otherwise Repeat Frames 13 and 14.

***Frame 15***: The sum of the number of neutrons and protons is equal to the atomic weight of an atom.

***Example***: Oxygen has 8 protons and 8 neutrons. Thus, the atomic weight of oxygen is 8 + 8 = 16.

Sodium atom has 11 protons and 12 neutrons. Check the statement below that is true. Atomic weight of sodium is:

(a) 11

(b) 12

(c) 23

(d) 1

If answer is correct go to Frame 16, otherwise Repeat Frame 15.

***Frame 16***: Nitrogen atom has 7 protons and 7 neutrons. The atomic weight of nitrogen is.... .

***Frame 17***: The atomic weight of hydrogen is 1. It has 1 proton, then the number of neutrons in hydrogen atom is......

If answer is correct go to Frame 18, otherwise Repeat Frames 15,16 and 17.

***Frame 18***: The number of shells in an atom may vary from 1 to 7, that is the number of shells in an atom cannot exceed 7, though it can be less than 7, i.e., (6, 5, 4, 3, 2, 1) depending upon the total number of electrons in a particular atom.

The number of electrons present in each shell is governed by 2 rules. *Rule 1*—The nth ($n$ is the number of shell) shell cannot have more than $2n^2$ n electrons.

***Example:***

For the first shell $n = 1$

therefore $2n^2 = 2.$ f $= 2$

It means 1st shell cannot have more than 2 electrons (if the atom has more electrons they will be in the other shells), though it can have less than 2 (if there are not enough electrons in a particular atom).

For the 2nd shell $n = 2$

Therefore, $2n^2 = 2 .2^2 = 2 .4 = 8$

It means 2nd shell cannot have more than 8 electrons though it can have less than 8 (as explained above).

According to this rule, there should not be more than:

(a) ..... electrons in 3rd shell.

(b) ...... electrons in 4th shell.

(c) ..... electrons in 5th shell.

(d) ..... electrons in 6th shell.

(e) ..... electrons in 7th shell.

If answers are correct go to Frame *19,* otherwise Repeat Frame 18.

***Frame 19*:** Oxygen has 8 electrons. According to the above mentioned rule write down the number of electrons in different shells.

Shell 1.......

Shell 2.......

Shell 3.......

If answers are correct go to Frame 20, otherwise Repeat Frames 18 and 19.

***Frame 20*:** The atomic number of chlorine is 17. Write down the number of electrons in different shells.

Shell 1.......

Shell 2.......

Shell 3.......

Shell 4.......

If answers are correct go to Frame 21, otherwise Repeat Frames 18, 19 and 20.

***Frame 21***: *Rule 2*—Last shell cannot have more than 8 electrons in an atom, and the last but one shell cannot have more than 18 electrons. Complete the following statements:

(a) An atom has 5 shells. The number of electrons in its 5th shell should not exceed......

(b) The number of electrons in its 4th shell should not exceed........

If answers are correct go to Frame 22, otherwise Repeat Frame 21.

***Frame 22:*** The atomic number of sodium atom is 11 and its atomic weight is 23. Write down the number of the following items of a sodium atom.

(a) Electrons...........

(b) Protons.........

(c) Protons and neutrons........

(d) Neutrons..........

(e) Electrons in 1st shell........

(f) Electrons in 2nd shell..........

(g) Electrons in 3rd shell...........

(h) Electrons in 4th shell...........

If answers are correct go to Frame 23, otherwise Repeat Frames 13-22.

***Frame 23:*** The atomic number of carbon is 6 and its atomic weight is 12. Complete the carbon atom by putting at proper places the number of electrons, protons and neutrons.

If answer is correct congratulations, you learned what we wanted to teach you through this programme. If not please repeat Frames 13-23.

## SOLUTIONS

***Frame 1***

Atom

***Frame 2***

(a) proton

(b) negative

(c) neutron

***Frame 3***

(a) positively charged

***Frame 4***

(c) neutral

***Frame 5***

(b) negatively charged

***Frame 6***

(b) neutron

***Frame 7***

(b) +4 units

***Frame 8***

X units

***Frame 9***

(c) protons and neutrons both

***Frame 10***

(b) B is a shell

***Frame 11***

(a) 40 protons

(b) will not be

***Frame 12***

(a) shells

(b) nucleus

(c) nucleus

***Frame 13***

(a) 11

(b) 11

***Frame 14***

(a) 8 protons

(b) 8 electrons

***Frame 15***

(c) 23

***Frame 16***

14

***Frame 17***

0

***Frame 18***

(a) 18

(b) 32

(c) 50

(d) 72

(e) 98

***Frame 19***

shell 1=2, shell 2=6, shell 3=0

***Frame 20***

shell 1=2, shell 2=8,

shell 3=7, shell 4=0

***Frame 21***

(a) 8

(b) 18

***Frame 22***

(a) 11

(b) 11

(c) 23

(d) 12

(e) 2

(f) 8

(g) 1

(h) 0

# 6

# Significance of Audio-Visual Devices

In an introductory class of audio-visual education from B.Ed. students it was asked some of their earliest school experiences. Following were some of their answers:

1. "The earliest school experience was when I made things out of plasticine."
2. "When in class 3rd we made a model of iglu out of sugar cubes."
3. "When I went to see a zoo in my 2nd grade."
4. "When we did gardening in a small corner of the school."
5. "When I played the role of little red riding hood in kindergarten."
6. "The puppet show I saw in the school."

Dozens of similar experiences were reported by these (college) students. All these are varied experiences but we see one similarity, they could remember the experiences where they are involved in doing something, when they were active, when more senses were involved. These are some of the experiences where effort was not made to memorise them, on the other hand we do not always remember the facts or concepts that we try hard to memorise. From this we can generalise that learning becomes comparatively permanent when concreteness is there in the experiences. The various experiences help in developing concepts.

The process of building concepts operates naturally from the time a child begins to draw certain conclusions from his experiences and applies these conclusions to new situations. It continues thereafter as he makes new generalisations from new experiences and experiences in which new and old experiences are combined.

The overall activity of building concepts, therefore is a realistic definition of Education. Now we can say that two elements are involved in building concepts: (1) Certain amount of concrete experiences, and (2) Combining and recombining these concrete experiences in many ways. When we apply this to classroom teaching, audio-visual aids play a very important role in concept formation and therefore in permanent learning.

## The Demonstrations

### *Kinds of Exhibits*

There is a big variety of exhibits. Sometimes they are three-dimensional working models, sometimes a series of photographs or photographs mixed with charts, models and real objects. But exhibits are essentially something one sees as a spectator. Usually one is not involved in handling any thing or working with the material. Science teachers take their students to show the science fairs where different kinds of exhibits can be seen. They are also of great educational value. Sometimes they may also influence the attitudes.

When students are involved in making exhibits, it becomes a direct experience for them. Investigatory science projects done by students can be displayed as exhibits. Exhibits can also be used by the science teachers to teach subject matter.

***Museum*:** Museums are places where items and exhibits of knowledge are assembled, protected and studied. In big cities there are public museums of various kinds like art museums, history museums, science museums, and natural history museums, etc. Here we are concerned with science and natural history museums. In our country there are not very many science and natural history museums. But if they are available in your cities, advantage should be taken from them.

Mostly Science and Natural History Museums (Science museum in Bangalore, Natural History Museum in Darjeeling and New Delhi) have two functions: (i) presentation and display of the materials, and (ii) to work with classroom teachers on specific curriculum unit. If this facility is given by the public museums, science teachers should take full advantage of such arrangements.

The other kind of museums are college and school museums for various subjects. Biology museums are quite common museums established in some form or the other in many schools and colleges. Museum should not be considered as a collection of various items and objects. It is more than that, it must be viewed with an idea, a process, and an objective. Therefore, science museums should not emphasise only on the product of science but due importance should also be given to science processes.

There is a big scope of displaying of specimens and objects collected by students. Models and projects made by students could also be displayed there. In fact development of museum can be taken as one of the science club activities.

## Small Screen and Moving Pictures

Next on the cone towards little more abstraction comes the TV and motion pictures because we are only the viewers of these audio-visual aids. First we will discuss the motion pictures.

### *Moving Pictures*

***16-mm Educational Films:*** It is not possible to learn everything by doing or first-hand viewing. We have to get some of our experiences indirectly, 16-mm films play a very important role in giving this indirect experience. Motion pictures present an abstracted version of the real events omitting unnecessary and unimportant details, they can, sometimes dramatised events so effectively filmed that we feel as though we are present at the reality itself. No other medium has brought so much information and scientific knowledge to the classrooms as the educational sound motion pictures.

There are various kinds of motion pictures. Here we are concerned with the *educational films*. Educational films include *documentary* and *instructional films*. Though some documentary films can be used in classroom teaching but they are especially planned for classroom teaching purposes. On the other hand *instructional films* are specially planned to achieve certain educational objectives and are made in specific subject areas for teaching purposes.

These films can be background films or direct teaching films. They help to promote to achieve a skill, an attitude or to convey certain facts, information, phenomenon or theory. As a science teacher we will be mainly concerned with the instructional films in our areas.

Though we are only spectators before a motion picture but the learning outcomes are quite effective because of certain specific values of motion pictures.

For 16-mm films, 16-mm movie projectors are needed. Now video cassettes of 16-mm films are available, which can be seen on TV. with V.C.P. or V.C.R.

***Strengths of Motion Pictures:***

1. Motion pictures motivate the students and compel attention.
2. Movement can very effectively be shown by motion pictures which is not possible in any other aid except direct experience.
3. It is an edited version of reality, the editing involves manipulation of time, space and by eliminating distractions; it gives relationships of things, ideas and

events that might well be overlooked in real life. For example, food web, food chain, etc., can be seen as continuous processes which is not possible in reality.

4. Time can be controlled in a motion picture with the help of slow motion photography, "movements in animals, insects, flight of birds" can be understood very easily. With the help of time-lapse photography it is possible to see the blooming of flowers, germination of seeds, growths of plants, etc.
5. Motion pictures can enlarge or reduce the actual size of the objects for better compensation. By means of microscopic lens attachments microscopic organisms like algae, fungi, protozoas, and blood circulation in capillaries, etc., can easily be seen and understood by students.
6. By X-ray cine photography techniques movements of internal organs of animals — action of arms, legs, movement of the heel, swallowing of food, heart-beating/ living movements, etc., can be shown.
7. Motion pictures can bring past and distant present into the classroom.
8. With the help of animation techniques motion pictures can show processes which cannot be seen by human eye even with the help of microscope or telescope. For example, behaviour of molecules in solids, liquids and gases, phenomenon of nuclear fission and fusion, etc.
9. Motion pictures can reproduce the record of events like lunar and solar eclipses, etc., which are not always possible to see. Complicated techniques can be filmed and reproduced when required.
10. Motion pictures promote understanding of abstract relationship, offer a satisfying aesthetic experience, influence and even change in attitudes.

With so many values, motion pictures not only expedite the rate of learning but they also increase the scope of teaching. This is an important consideration in the field of science where mass of learning and the learning processes are expanding so rapidly that it is to keep hand up with them. Motion pictures enable the teacher to clarify the difficult concepts in less time than with any other technique.

In order to get the most effective use of films or video tapes, the following necessary information and skills should be acquired:

1. Where to obtain appropriate films or video tapes?
2. How to use such films and video tapes in your teaching?
3. How to operate the 16-mm film projector, V.C.R, V.C.R.?

School, college and university teachers have to spend some time to find films or video cassettes which they can integrate into their courses. It is possible to find suitable films or video cassettes on various science subjects. Securing catalogues will be important for this purpose.

For an effective use of films or video cassettes, preparation and follow-up are two very important steps to be taken by a teacher. Teacher should select the film or video cassette which goes along with the topic. The film or video cassette should be previewed and important points to be covered in the lesson should be outlined.

It will be advisable to frame few short answer questions. These questions can be given to the students before showing the film or video. During the follow-up activity unanswered questions and other important points should be discussed. Such activities help in coordinating the film or video with the topic. In the absence of such coordination sometimes the film or video becomes a wastage of time and effort.

The next point is to learn to operate the 16-mm projector and V.C.P./V.C.R. which is not very difficult to learn. Ordinarily only a few hours instruction followed by a brief practice results in satisfactory operation. Simple threading and operating instructions are given somewhere on the projector, which should be followed carefully. The possibilities of error should not discourage the teacher but it should be taken as challenge. The thrill of this educational device will compensate for minor disappointments.

### *Caution to be Observed in Using Films or Videos*

Motion pictures may not necessarily be effective for teaching everything. As a teacher you have to be cautious about certain things.

- You have to think about the *effectiveness* of the film or video. If other experiences like direct or contrived are more meaningful, film or video should not be used.
- Sometimes the film or video gives a wrong idea about the *time notion* and *size notion*. It should be clarified to the students by comparing it with a known object or by giving a familiar example.
- If inexpensive devices are available with the *same effectiveness* as motion pictures, then less expensive devices should be used.
- Teacher should see that the film or video is adequate for the comprehension of his/her students.

### *Small Screen*

***School T.V. Programmes:*** Television has all the potentialities of motion picture with a little more concreteness because of the "on the spot coverage" and the nature of the TV programmes, as most of them are specially produced for a particular audience. This is more so for educational TV programmes. In many countries educational TV programmes include series of science programmes.

### *Educational TV in India*

Experimental TV Service was inaugurated on September 15th, 1959. This programme was planned for educational and cultural values of community and designed for community viewing. In the meantime Ford Foundation in India was approached to assist in the development of educational TV. Ford Foundation team of TV experts visited India in February 1960. In 1961 Television was started in India as an educational TV project. The project was launched in close collaboration with the Delhi Directorate of Education and with the financial assistance of the Ford Foundation in 1961.

The educational TV programmes were integrated with the school syllabus based on the principle of direct teaching. Direct teaching implied that the TV programmes were directly related with the school syllabus. They were proposed to help in classroom teaching by bringing into classroom additional resources through the medium of TV. The students were benefited from talented teachers, best and expensive equipment and other audio-visual aids which were not usually available in schools.

In the project the teachers were also fully involved from the planning to evaluation. This arrangement of instructional TV was working well in Delhi schools because there was a set prescribed syllabus. In some other advanced countries there is not any set syllabus for all but they have different syllabi choosen by teachers and students. Their educational TV programmes are not strictly integrated with the syllabi.

In Delhi the educational programmes started in General Science, Social Studies, Physics and Chemistry. Later Biology, Maths and Geography were also included.

For proper functioning of educational TV and better coordination between Doordarshan and Schools a Television Branch was set up in 1967 in Delhi Directorate of Education. The TV branch used to divide the syllabus term-wise and week-wise, and select lessons for TV teaching. To ensure close relationship between TV and classroom teaching, a package of a uniform timetable, set examination days and working hours, and a uniform weekly syllabus in the form of a TV booklet was to be distributed to all viewing schools in the beginning of the school session. Such TV booklets used to be prepared for each subject taught on TV. But now we do not have such school TV programmes in Delhi or elsewhere in the country.

This kind of educational TV programme if started again can solve several problems of science teaching in our schools:

***Merits:***

1. It solves the problem of unequipped laboratories to certain extent. Students are able to see effective demonstrations and expensive apparatus which is otherwise not possible for them to see.
2. TV teachers prepare and rehearse a lesson, spend enough time to collect relevant material and up-to-date information.
3. It has all advantages of films plus being a live presentation seems more natural. Only relevant portions of films are shown, along with teacher's explanation and other visual material.

***Demerits:***

1. It is a one way process and students are passive viewers.
2. There is no opportunity for an active participation of the learner during TV lesson.

### *Better Utilisation of Educational TV*

The disadvantages can be removed if some preparation is done and precautions are taken by the classroom teacher.

For effective use of any mass media three things are important: proper reception, proper seating arrangement, carefully planned preparation and follow-up.

Good reception is one of the basic important things for effective viewing. The television branch in Delhi had the responsibility of installing and maintenance of TV sets. The coordination had to be done by the teachers and administration of the school. Classroom teacher was also be familiar with proper handling and minor adjustments of a TV set.

Next important thing is the proper seating arrangements. If a big crowd is sitting on the floor (without any discipline) in an auditorium or classroom, not sure of proper viewing, this kind of arrangement will be a waste of time. The classroom teacher was to see that there was proper seating arrangement.

TV presentation is only one part of the learning procedure. Preparation for the TV lesson or the pre-telecast activities and the follow-up or post-telecast activities by the classroom teacher are equally important without which the TV lesson may not be very much meaningful.

Pre-telecast activities included the preparation by the subject teacher. Preparation varied with class to class and also depended on the topic, the classroom teacher could see the topic of the TV lesson in advance from the TV booklet. During pre-telecast activities his job was to make the students ready to receive the TV instructions. He could ask motivational questions. If some background knowledge was needed it was to be given at this time. The timetable was arranged in such a way that there were about 10 minutes before and after the actual telecast, usually the presentation was of 20 minutes duration, and class periods varied from 35-40 minutes.

After the TV presentation follow-up activities included clarification of points of doubts raised by students. Recapitulation questions could also be asked to see how much they have grasped. With little encouragement by the classroom teacher many activities could be followed by students as continuation of the TV presentation for better understanding.

The TV branch of Delhi also tried to evaluate the lessons. They used to send a printed pad of evaluation check sheets to each school for the constant feedback of TV lessons. If various people connected with educational TV production do their jobs consciously the E.T.V. programmes can be quite effective. You as a science teacher has to play an important role.

During your practice teaching you can observe some science lessons on video tapes, evaluate them according to the evaluation check sheet developed by you and give suggestions for further improvement.

## Pictures and Radio Recordings

Next stage on the cone of experiences after motion pictures is the number of those devices which can be called as one-dimensional aids. We will take the scope of each in teaching science one by one.

### *The Pictures*

***Kinds of Still Pictures:*** Various kinds of still pictures can be used in teaching science. They include photographs, illustrations and slides. These visual materials are of great importance in teaching learning situations.

Photographs and illustrations can be used without projection, slides can be projected on the screen with slide projectors. Therefore, pictures can be divided into two kinds: (i) unprojected, and (ii) projected.

***Unprojected Pictures:*** Photographs, illustrations and clippings of photographs from magazines and newspapers can be collected for teaching purposes. Pictures give a correct impression of the object or situation and motivate and enrich teaching. It is good to have relevant pictures on the walls of science rooms — pictures of great scientists and some of their apparatus, etc. Pictures can also be used in teaching for motivation, introduction, presentation or recapitulation as the need may be.

Pictures and illustrations with some theme can be displayed on a bulletin board.

### *Bulletin Board*

It is a place for posting "bulletins", as the name suggests. Bulletin boards can be used in many different ways and as an effective "teaching device". Bulletin boards inside and outside the science rooms and laboratories can be used for putting visual materials (pictures, illustrations) with relevant heading and some labelling for supplementing the teaching.

It should be of scientific interest and importance. With proper teacher guidance and motivation bulletin boards can provide opportunity of developing creativity, responsibility and interest among the students. Other than displaying the photographs and illustrations, bulletin boards can also be used for displaying reports of science projects done by students, science news and cuttings from science magazines, collected and presented by them.

### *Projected Pictures*

Projection of visuals have certain advantages. For example, projection magnifies the material, therefore, a bigger group can easily see it. It is somehow motivating, therefore, compels attention.

Different kinds of projectors aroused in classrooms for projecting still pictures. They all are very easy to use, with little planning on the part of the teacher, teaching can be quite effective.

### *Opaque Projector of Episcope*

The opaque projector projects and simultaneously enlarges material directly from original books, magazines, etc. All kinds of written, printed or pictorial matter in any sequence can be projected for teaching purpose. It can also project opaque thin objects like leaves, shells, sample of fabrics, butterflies, moths, etc.

Designs for posters, maps, pictures for displaying on bulletin boards etc., can be projected in the desired enlarged forms and then traced. The opaque projector though has many unexplored possibilities but is not being used much because of its bulk, weight and requirement of a very dark room.

***Slide Projector***

Such projectors usually project 35 mm slides. It is essentially a simple machine, easy to operate, inexpensive and light weight. It has most of the points of practical importance.

Slides can be shown individually in any desired order.

## Significance of Slides

1. Slides are of great value in visual teaching situation, when motion is of little or no importance, for example, in teaching structural details.
2. Great variety of visual material such as pictures, charts, graphs, diagrams, maps, tables, anything that can be photographed can be put on a slide.
3. Slides are available in black and white and coloured forms on many science subjects covering a wide range of topics and levels.
4. Slides are comparatively inexpensive.
5. Slides require only a slight darkening of the room.
6. In some little more expensive projectors the slides can be moved automatically either from a master control or from a synchronising device on a sound track. This arrangement of slide-tape programme is an effective device as the well prepared commentary makes it more meaningful.

## Slides' Effective Utilisation

The use of this visual aid will be more meaningful if following points are considered while using them:

1. The science teacher has to explore if suitable slides are available on the particular topics.
2. Second step is the careful preview. The preview helps in deciding on how to use the material to the best advantage. Is it stimulating enough for introduction on the whole

lesson could be developed with it or only suited for summarising the lesson.

3. Teacher has to think about the purpose also. If showing of motion is necessary for understanding, then slide is not the best medium.
4. If teaching involves a series of step by step development, one leading to the next logical sequence the slides are well suited.

Showing slides is only one part of the process. Good discussion and further activities could be developed from it.

### *The Micro Projector*

One of the more specialised types of projection equipment is the micro projector. As its name indicates it shows enlarged images of stained sections of microscopic slides or other biological material mounted on microscopic slides so that the bigger group can see it.

The advantages of micro projector are that:

1. It presents an enlarged picture of the object on the slide for common view.
2. It assures the teacher that his students are seeing precisely what he wants them to see.

Some manufacturers make low power projection attachments for slide projectors. Some are simple extensions of a microscope using a more powerful illumination system with a small view screen. This can be easily shown in small groups. But the full scale micro-projectors are quite expensive.

Projectors of any kind are a boon to classroom teachers. These technological tools, if coordinated wisely make the classroom teaching quite effective, because of the novelty of presentation, students become motivated and interested.

## Recordings and Radio

### *Language and Voice Modulation*

These are primarily auditory aids. If the speaker on radio or recordings uses simple everyday language with meaningful associations with familiar examples the listening can become quite concrete. On the other hand if the language is difficult without any associations, it could be an abstract thing for listeners. Being only an auditory aid language and voice modulation, etc., play a very important role in the effectiveness of these programmes.

***Values of Radio***

Immediately we are able to get the news of important events from all over the world, which is not possible from any other media except video or TV. Many events are of great importance for science classes, some of which are mentioned below:

1. Radio brings reality to the classroom The spectator who describes the events with his tone and description can brings sense of participation amongst listeners.
2. Radio programmes if well rehearsed and well presented in the form of stories bring the personal feeling of the actor, different sound effects and voices make them more realistic and meaningful.
3. It brings variety to the classroom and helps in teaching
4. Though in a well prepared programme enough expenses may be involved but radio being a mass media technology, on the whole it is inexpensive.

***Limitations:***

1. It is still a one way communication
2. Availability and maintenance of radio set may be a problem.
3. Radio broadcast timings always do not coincide with the class timings. Provision of radio broadcasts in the school time table is made only in a very few schools.

***Educational Radio Programmes in Our Country***

Educational broadcasts for schools in the AIR were regular features from the last few decades. These programmes included science lessons for students and teachers at various levels. Teachers' programmes were on content as well as on methodology. Before the board examinations revision lessons on various science subjects were quite popular. These were used to be given by experienced science teachers and educators.

The school broadcast was a separate unit of AIR in New Delhi. The unit tried to coordinate between various agencies to improve the effectiveness of the broadcasts. Meetings were held where people were invited from Directorate of Education Delhi, Municipal Corporation Delhi, New Delhi Municipal Council and Delhi University to plan future programmes and give suggestions for their improvement.

Radio broadcast schedule in the form of a chart was printed and sent to the schools in the beginning of the year. Teachers could utilise the relevant broadcast in their classes. Radios and transistors were or could be easily available in the schools. If the teachers are conscious of the radio programmes students can be benefited Pre- and post-broadcast activities improved the effectiveness of the programmes as discussed in TV programmes.

### *Recordings*

All the values of radio can be applied to recordings also. But the disadvantages can be eliminated like timing, administration and one way communication problems. Recordings from many sources can be utilised in the classrooms, like spool-tape records and cassette records.

***Advantages of Recordings:***

1. Recordings can become two way communication as it can be stopped, discussed and replayed wherever it is necessary, it can be controlled by us.
2. Recordings on cassette or spool-tape-recorder can be made in the school also. A commentary can be written and recorded on the available slides in your subject in the school for effective use and reuse.

3. Good radio programmes can be recorded and replayed in class whenever needed.
4. Scientific talks by eminent personalities can be taped and reproduced in the class.

## Sighted Symbols

### *Kinds of Visual Symbols*

They include the abstract representation of real pictures. Various kinds of visual symbols used in communication are drawings, sketches, diagrams, graphs, cartoons, flat maps, etc. They all can be projected on screen, and can also be shown on the blackboard or on the overhead projector.

Diagrams and charts, etc., are part of everyday work for most of the science teachers. Almost all the science teachers use blackboard to show in the structure and happenings of various things in their subjects.

Commercially made charts are available in various subjects. Charts and other graphic aids are also made by various Governmental agencies. Central Institute of Educational Technology (CIET) of NCERT New Delhi develops various kinds of graphic aids for school purposes. Charts are also made in schools by teachers and students and can be utilised in teaching science quite effectively.

### *Chalkboard*

Blackboard now is usually called a chalkboard because of variations in colours. It is one of the very important but quite neglected aid of the classrooms. If it is properly used can become a valuable aid in teaching science.

It should not be taken for granted that you already know the use of chalkboard, but the techniques should be learned and practised especially as a trainee teacher.

*Importance of Chalkboard:*

1. This aid is used with all other aids for better clarification, for example, the main points of a demonstration planning

and summarising of a field trip, questions for a motion picture and so on can be written on the board.

2. It gives a concrete form to abstract and vague statements and ideas.
3. It helps in developing the skill of drawing a particular diagram.
4. It is less time consuming and comparatively cheaper.
5. It can be used for some students activities like quizes, competitions and discussions, etc.
6. Some chalkboards can be used for permanent outlines of maps, graphs, tables and other special materials where some filling up can serve the purpose.
7. If properly used chalkboard becomes an attractive point and holds students attention.

***Effective Use of Chalkboard:***

1. Clean the board properly before using it.
2. Try to write neat and bold so that it is legible.
3. Do not write too much on the board. Start writing from top left hand. Give headings and sub-headings in the written summary.
4. If diagrams are used for referring or introduction, they can be made beforehand. If you want the students to draw the diagrams then it should be drawn in front of them and they should be asked to draw side by side. In the same way a complicated process should not be drawn beforehand, but it should be developed before them for clarification and better understanding.
5. If necessary, some tools can be used to draw on the board.
6. Coloured chalks should be used for making complex diagrams.

### *Overhead Projector*

Any thing which is written or drawn on a chalkboard can also be projected by an overhead projector with some additional advantages.

An overhead projector projects material overhead or over shoulder the teacher using it on to a screen behind and above him. They generally have 25 cm x 25 cm writing area which is big enough for writing. What is required to be shown on the screen is written on a glass plate covered with sheet or roll of transparent plastic. Variety of fabrics tipped pens or china marking pencils are used for writing and drawing. Wet cloth can be used for erasing and reusing the sheets.

***Advantages of Overhead Projector:***

1. It is very simple and convenient to use the machine.
2. "Ibis projector unlike any other projectors is operated from the front of the room and the teacher faces class as he or she writes or points on already made transparencies.
3. Plastic roll attachment is quite useful. Diagrams, points of the lesson, assignments, tests and similar material can be effectively presented gradually with a minimum of time and effort.
4. An important feature of overhead projector is the provision of "overlays" or 'build ups," which have successive layers of transparencies in black and white coloured or both showing cumulative stages of development, sequences and sectional views, which is quite effective in many scientific processes and phenomenon.

***Limitations:***

1. Though it is not very heavy but a voluminous projector.
2. It is quite expensive and needs some planning.

## Oral Symbols

This is the most abstract and one of the most important forms of experience we get and give to our students. Our job as a teacher is to make sure that the various terms, ideas, words, principles and other abstractions used should be meaningful for students. The ideas and concepts which can be translated into verbal symbols come by getting other concrete experiences. Rich experiences gained by students help them to understand the verbal symbols meaningfully and therefore they can read and understand the textbooks also. Ability to speak or read the word does not mean that one also understands the object, process or phenomenon.

If from the very beginning science is taught by reading books or abstraction only, students usually acquire the habit of memorising and accepting verbal formulation given by others. This is no way of teaching science.

Various kind of experiences discussed on the core of experience from direct purposeful experience to visual symbols, all help in the development of verbal symbols meaningfully. Then reading is not only the process of reproduction of verbal symbols but also involves a thinking process, which includes meaning of words.

Textbooks and other reference books provide the most to the students, because all teaching is not possible by any one particular aid or experience. But the textbooks should be written in understandable vocabulary of that level. Sometimes in primary classes the language of science textbooks is much more difficult than the language in the language textbooks of that particular class. Another important aspect is also the style of writing and format. In writing and specially in teaching if references and examples are given from their environment and known things learning will be more meaningful.

## Competent Utilisation of Audio-Visual Supports

A variety of teaching aids has been discussed in this section with some details. It is difficult to select few aids and to say that they are more important than others, or this should be used more often than others, etc. The value of an aid does not depend wholly on its quality itself but it also greatly depends upon the way it is used and the particular time it is used.

The qualities of various aids have also been discussed. No matter how perfect the aid may be, if it is irrelevant to the occasion of its use then it is of little value and an ordinary aid with not too many qualities becomes indispensable if it is used relevantly at the right time. For example, there is film which shows the animals under the sea or life in a desert in relation to their ecology. Neither the materials nor this relationship could be shown in a class by any other media or field trip as effectively as the film could. But always it should not be used as a substitute for other materials of the classroom.

## The Significance

If we see teaching of science in schools of our country by and large we will find teachers lecturing or even reading out from the books and explaining few things on the blackboard. Some teachers use some demonstrations, charts and models but not very frequently and many a times students memorise things without understanding. Audio-visual aids if properly used help in teaching learning process in many ways (as given below) and can ensure quick and effective learning:

1. The timely use of proper aids compels attention, develops interest and motivates students.
2. They break the monotony of teacher's talk, reduce verbalism, save time (of long verbal explanations) and give a better idea of the real things.
3. Audio-visual aids can make learning experiences far more concrete; therefore clarification of concepts, better understanding and long lasting learning are possible.
4. Great many teaching problems can be solved partly or wholly by the proper use of rich experiences (through audio-visual aids); therefore they offer great opportunities for improved learning.

***The Kind:***

Variety of audio-visual aids can be used in science teaching. They can be classified in various ways.

(a) One way of grouping them is as follows:

*Visual Aids* — charts, photographs, diagrams, static and working models, etc.

*Auditory Aids* — radio, recordings on tapes, and cassettes.

*Audio-visual Aids* — films, T.V., etc.

(b) Another way of classifying can be:

*Graphic Aids* — diagrams, photographs, charts, play cards.

*Three-Dimensional Aids* — models, specimens, real objects, apparatus, dioramas, etc.

*Projected Aids* — slides, films, etc.

*Aids Through Activity* — excursions, projects setting and maintenance of aquarium, vivarium, botanical garden, etc.

## Experiences' Edger Dale Cone

Some of these aids are more concrete in nature and some of them are comparatively more abstract. Edger Dale has arranged the various audio-visual aids in pictorial form which he called "Cone of Experiences".

This is a kind of visual aid to visualise and explain the interrelationship of various types of audio-visual material as well as their position in learning process.

At the top of the Cone are verbal symbols which are most abstract and at the base are the direct purposeful experiences which are most concrete form of experiences. This is only a pictorial form where all sorts of aids and experiences are arranged in a Cone. The various bands of the cone representing various experiences and aids, should not be considered as rigid divisions or watertight compartments. They overlap and blend with each other. For example you can be a viewer of an exhibit or a person who made it, you can observe a demonstration or can demonstrate it yourself. It is also not being suggested that the various aids and materials are arranged in the form of their effectiveness but they are arranged from most abstract to most concrete.

As a teacher we should also know that abstract ideas, concepts, or generalisations are not possible without rich meaningful concrete experiences. A science teacher has to pick and choose the aids according to the maturity levels of the students and the topic to be studied. We will discuss the strengths and limitations of each in teaching sciences.

Verbal symbols
↓
Visual symbols
↓
Radio Recordings
↓
Still pictures
↓
Motion pictures
↓
Exhibits
↓
Field trips
↓
Demonstrations
↓
Dramatic participation
↓
Contrived experiences
↓
Real Direct experiences

**Fig.** *Edger Dale Cone of Experiences*

## Straight Determined Experiences

The base of the Cone of experiences represents "direct purposeful experiences." Here the word 'purposeful' is very important. Every direct experience may not be very meaningful, therefore, it may not be purposeful. We have to see the direct experience in terms to learning outcome. In cases where the real things are too small or too big to comprehend, direct experiences are not very effective and therefore are not purposeful. For example, "structure of atom," "working of a factory," "water supply in a city" etc., can be understood better by some indirect experiences like models, maps or charts, etc.

In teaching science many direct experiences can be given to the students for effective comprehension. Some examples are: observing real flowers, leaves, plants, insects; dissecting animals; taking a walk through woods; going to the seashore and observing marine animals; doing salt analysis in the lab; setting and maintaining an aquarium, etc. In such cases learning is by direct participation.

The science teacher has to decide what direct experiences will be purposeful for his classes and then try to give them as many experiences as possible because direct and concrete experiences soon become associated with abstractions and help in developing more difficult concepts.

## Strategic Experiences

### *What are Contrived Experiences?*

Next in the hierarchy towards abstraction are contrived experiences. Examples of contrived experiences are static models, working models, specimens, dioramas, etc. Contrived experiences may differ from the original in size (big things are made smaller and smaller things are made bigger) and complexity. It usually is a simplified and edited version of the real thing, where the unnecessary details can be removed to make the learning clear.

For example a petroleum refinery is difficult to comprehend in a real situation but its model will be more meaningful. In the same way models of places where it is difficult or impossible to reach and see can give a clear idea, like models of globe, volcano, parts and systems of the body, bottom of the ocean, view of a forest, polar region, etc. Sometimes we imitate the whole natural habitat in the form of dioramas and keep in museums. Cut-away or half cut models are extremely useful in teaching internal structure of eye, ear, stem, root and automobile, etc.

There are some examples where nobody has seen the real thing, the models are made on the basis of indirect evidences like model of atom, DNA structure. Here the whole concept is developed on the basis of these imaginary models.

While teaching through the models, teacher should give the idea of the real thing, regarding their size and complexity. All models are not correct reproduction of their originals, they are only simplified versions.

### *Objects and Specimens*

Objects and specimens are very common in science laboratories. These are also examples of contrived experiences. We collect rocks from various places. Different kinds of plants and animals are collected; pressing, preserving and sniffing are done for storing and study purposes. The objects and specimens are taken from the real settings. They are samples of real things minus real settings. The specimens are collected and stored so that they are readily available for study purposes. Another most important advantage of objects and specimens which is not otherwise possible in direct experience is that they can be arranged into groups and classes.

## Exaggerated Involvement

This has been placed on the 3rd band of Cone of experiences. Dramatisation means substitute for real experience of reconstruction of the original reality. There are many things we cannot possibly experience at first hand. There is a great value of dramatisation in education. Students can participate in a dramatisation or watch some kind of dramatisation. Both are valuable experiences but participation is much more meaningful and closer to reality than only watching.

The question is "What is the scope of dramatisation in science teaching?" Dramatic acts are quite popular in languages and social sciences. In sciences also the scope is not limited.

The films made on the work of various scientists are only possible because of dramatisation. Such films are quite effective as such experiences are otherwise not possible.

Students of primary and middle classes participate in scientific dramatised act in schools, such activities are also brought to science fairs, etc. Some very abstract and uninteresting ideas are taken for dramatisations, for example, different students act as various components of solar system with proper costumes, dialogues, songs, music and dance, and the abstract concepts become clear and leave a long lasting impact on participants and viewers; students act as various petroleum products and explain how they are formed and utilised like coal, petrol, vaseline, synthetic rubber and plastic, etc. Functions of vitamins and other components of food can also be taught through dramatisation.

It is also possible to use dramatisation in classroom teaching where costumes are not necessarily required but different students can remember their parts and act out in the classroom.

Interdependence of various components of an ecosystem and balance of nature can also be explained interestingly through dramatisation. An imaginative science teacher can think of many such topics which can be taught more effectively through such activities. Thus, dramatisation can become an effective teaching aid in teaching science.

## The Exhibition

Demonstrations are quite familiar activities in science class-rooms. Their merits and demerits have been discussed at length in the Cone of experiences they have been placed on the fourth band from the base because it is essentially a process of observing. It differs from the first three bands, which are essentially doing. Demonstrations are used to show how something is done or not done. When demonstration is followed by doing on the part of students, it becomes very much meaningful.

Demonstrations are used to clarify ideas, and help to develop skills, processes and attitudes. They are not limited only to demonstrations through apparatus only. They can be used to clarify abstract ideas on the chalkboard, through slides or motion picture also.

Demonstrations can be improved to a great extent if certain points are kept in mind while planning and doing a demonstration.

- Plan all steps of demonstration in advance.
  - — Rehearse the demonstration before going to the classroom.
  - — Keep the demonstration simple as far as possible.
  - — Keep the students involved.
- Make it sure that all the students can see it.
  - — Outline various points on the board.
  - — Keep summarising various steps.

## Ground Visits or Expedition

### *What is Field Trip or Excursion?*

A field trip or excursion is a planned visit to a point outside the regular classroom. It may be in the school, out in the community or it could be a long trip to far away places. Usually in field trips to places like visits to a factory, observatory, agricultural institute, poultry farms, museums, etc., we often see other people doing things. As spectators we are not involved but we directly watch it and get a first hand knowledge. Therefore, field trip is an excellent bridge between the work of the classroom and the work of the outside world. The chief difference between a field trip and other educational experiences is that the student get their experiences in the field and not in the classroom. Because of its nature to be mostly observation it is kept on the 5th band of the Cone of experiences as less concrete than other experiences discussed before.

But if the tield trips are planned and arranged in such a way that they go beyond observation, for example of a sea-beach, on a pond where they can also touch, feel and collect things, it becomes a *direct experience* for them. Such a variation in the field trip indicates again how the bands of the Cone interlap and blend into one another.

### *Importance of the Field Trips*

Excursions or Field Trips are of great educational value especially in science subjects. The classroom is a limited place, bounded normally by four walls and meagrely equipped for the task or providing students with worthwhile experiences. The environment outside the classroom has no bounds; it has almost every conceivable situation that a teacher might wish to utilise. In school corridor students may study writing system for supply of electricity to different rooms and laboratories, may determine the power (in watts). On school grounds there may be various types of plants, birds, insects, different kinds of soil, sunshine and shadows, building materials, bicycles, scooters and cars. And just beyond the school boundaries lie the unlimited resources of community.

### *Contributions of an Excursion or Field Trip*

1. Field experiences are first hand experiences. They arise from direct learning situations. Sometimes they play the same role or even better in the learning of science as do experiments and demonstrations.
2. Field experiences tend to be much more meaningful and permit easier transfer of learning to solutions of real life problems.
3. Fieldwork if properly organised awakens many interests that classroom work cannot arouse. Fieldwork is the study of actual objects which stimulate more curiosity than to ideas. Out of almost any situation encountered in the field can develop into some challenging problems.
4. Fieldwork permits first-hand study of many things that cannot be brought into the classroom because of size and

other inconvenience, e.g., it is only outside the class that the students can be acquainted with the flora and fauna of the area.

5. Fieldwork permits a class to engage in activities that are too noisy or too violent to be used in the classroom, e.g., a model airplane, gasoline engine if demonstrated in the classroom would disturb the other neighbouring classes too.
6. Outdoors, students are able to work with large size materials, e.g., an iron piece pipe makes a more impressive lever than a routine stick and erosion is better demonstrated from a garden hose than with a tiny trickle from a yet drawn in a glass tube.

### *Some Special Experiences of Field Trips*

Fieldwork brings students in contact with many objects. The observation and manipulation of objects in the environment are bound to arise questions. Attempts to answer questions give rise to new problems.

Field trips can be used for review and drill. Ideas learnt in classroom can be better fixed in mind in actual situations, e.g., by visiting factories the ideas of running a plant and the products manufactured can be better fixed in students' minds, or visiting a zoological park or a botanical garden, more abstract ideas about animal and plant life can be more clarified to students.

### *Difficulties Experienced in Arranging Field Trips*

Lack of availability of good inventories regarding field experiences, school policies, and transportation problems create hurdles in arranging field trips. Therefore, the teacher will have to take a key role if he wants to arrange the trip.

#### *Procedure to Arrange Field Trips:*

1. Survey of the place of excursion should be taken before going to the field trip, so that the teacher knows

beforehand what their students are to see there; or what relevant literature should be studied if places are far away.

2. Objectives of the field trips should be very carefully identified to make a field trip a success. If this is done teacher knows what he is going to teach and the students know what they are going to learn there.
3. Permission from the authorities and parents of students should be taken well in advance.
4. Appropriate activities compatible to the identified objectives should be listed and given as well as discussed with the students beforehand.
5. If there is a need for transport and place to stay they should be arranged in advance.

***Preparation of Students/or Excursion:***

1. It is essential to brief students after arriving at the area. This should be as minimum as possible. A long lecture may deprive the students of an opportunity to explore the place.
2. Students should be divided into small groups appointing a group leader for each group. Group leaders may help in running fieldwork smoothly and quickly.
3. You as a teacher should be aware of the fact that there are slow as well as rapid learners. Activities and responsibilities should be so divided among the students as to be equally shared by both type of students without any feeling that they are slow learners or rapid learners.
4. Students should be made aware that there may be some difficulties like: (a) noise while visiting any working plant in a factory, (b) listening to the guide due to distance while standing around him, or (c) technical language used in explaining particular process; and they should try to solve such difficulties themselves as far as possible.

5. Students may be advised to take notes and draw diagrams, whenever they think it is necessary.

***Teacher's Role:***

The teacher's role is very important for a successful field trip. Some of them are listed below:

1. Watches students closely and gives specific help as and when needed.
2. Recognises student's achievement.
3. Avoids frequent interruption in student's work but occasionally if it is justified.
4. Does not lecture while in field ?
5. Avoids loud voiced comments.
6. Acts as a guide, resource person or consultant.
7. Does the relevant follow-up activities of the trip ?

***Follow-up Activities:***

Follow-up activities are very important in any kind of field trip, it may be a short trip just outside the classroom or a long excursion to various far away places.

Follow-up activities vary according to the nature of excursion. If the field trip is taken just outside the classroom in the school lawns to find the population densities of various species of plants, the follow-up activity will include the pooling up of data collected by each student or each group of students and discussion on the result. If the field trip is an observational trip to places like factories, mills, observatories or hospitals, discussion about what they have learned is important. Field trip can also be taken to study the fauna and flora and collection of specimens from seashore, hill stations or any other places.

If such field trip is taken by the students follow-up activities will include pressing of plants, preservation of animals, drying of insects, classification and displaying of material. If proper care has not been taken just after or during the trip (in case of long trips) the materials collected cannot be utilised properly and goes to waste. Sometimes the students and teachers are quite enthusiastic while planning and taking the field trip, but forget about the material after the big excitement or do not do the follow-up activities properly, therefore, the field trip is not as effective as it should be.

In short we can say that follow-up activities should be planned according to the objectives of the field trip. Science teachers should plan their objectives for the field trip well and do the follow-up accordingly for better understanding.

7

# Curriculum Development

In modern educational system the concept of curriculum is that it should cater for the all-round development of the learners. The Dictionary of Education defined curriculum a body of prescribed educative experiences under school supervision, designed to provide an individual with the best possible training and experience to fit him for the society of which he is a part or to fit him for a trade or profession. In fact, according to the current concept, the curriculum includes the totality of experiences that the learners receive through all types of activities in and outside the classroom. The task of curriculum construction involves the selection of right experiences for the learners which will lead to the attainment of the educational objectives. It includes not only the course content, resources and materials for the purpose, but also the guidelines on the methods and approaches to the teaching of the subject, which makes the whole exercise easier.

Since the imbalance between the growth of science and the attitude of welfare for mankind has engaged the attention of the western thinkers, the Commission has rightly pointed out that "knowledge and wisdom, power and compassion, are out of balance." Hence, curriculum planners ought to note the warning seriously for future guidance, otherwise, it would not prove to be successful.

In modern terms of education, the term curriculum is commonly understood to mean a course of studies, because the original Latin word *(currere)* meant a course run to reach a certain goal. The earlier accepted aim of education was also confined to the acquisition of knowledge only and, as such, curriculum was identified with the items of knowledge imported through instruction in the institutions. But the modern idea of education has undergone a revolutionary change and its aims are now much wider in concepts, meaning and functions. The modern world is a scientific world and the modern civilization is a scientific civilization.

## Modern Scientific Development

The modern scientific developments have made the world much smaller. Events in today's world are happening much faster and the society is also changing rapidly with time. In such a context, it is imperative that the concepts and meaning of education will change with the passage of time. Education is now considered a most dynamic social process and the concept of curriculum as a whole has also acquired a wider connotation.

According to Cunningham, "curriculum is the tool in the hands of the artist to mould his material according to his ideals in his studio." Thus, curriculum is the instrument in the hands of the teacher, the artist to give shape to the materials, the students. The traditional concept of equating curriculum with subject knowledge is no longer tenable. Science curriculum, as a constituent of the total educational curriculum, must contribute to the attainment of the objectives of education.

As above mentioned, curriculum in the wider sense is overall experience of the learners through a variety of activities pursued inside the classroom, in the laboratory, on the playground or through the formal and informal association with the teaching community. Such experience, duly guided, helps in developing and enriching the personality of the learner.

And one of the major aims of education, as we all know, is the all-round development of the learners. Curriculum evolves from life itself and, as such, curriculum planning should be life-centred. A curriculum must also conform to the needs of the state and the society. At the same time it should be flexible and dynamic since our modern society is fast changing with time.

The curriculum should give more importance to the immediate needs of the individuals as well as the society and provide for their future needs too. We know that the psychological developments in a child are not abrupt but a continuous process. Therefore, whenever we are concerned with curriculum construction for any particular stage, we must see that it is continuous with the previous stage and is linked to the next stage.

The young learners should be given continuous, connected and sequential science experiences from the primary stage to the high or higher secondary stage of school education. Care should be taken in organising the course content so that the learners understand the basic scientific concepts, principles and definitions and see their application in everyday life.

The methods and approaches for imparting science education should also develop in the learners the power of observation, critical thinking, and scientific attitude. The approaches should be inbuilt in the curriculum structure to lead the learners through instruction and experiences, to develop the right interest and appreciation, and also the skill and ability to apply their knowledge of science in everyday life situations.

The science contents as well as the approaches should enable the learners to use their knowledge in solving their individual as well as social problems. A curriculum should be flexible and broad-based to change according to the needs of the child and the society. Since the exposure a learner is subjected to, in or outside the school, ultimately has to benefit the society at large, curriculum should therefore be child-centred as well as community-centred.

The learner has to live and move in a society and as such the curriculum must be related to his social environment. The curriculum on any subject should enable the learner to distinguish between right and wrong, good and bad. Thus, while preserving the best in his traditional and culture it must also enrich the cultural level of the society and the social environment. A well-planned curriculum helps the learner become socially adjusted in the society. A curriculum is not just a list of theoretical statements. It must be activity-centred to be effective and interesting to the learners and must encourage creativity. It must satisfy the specific ability of each individual learner and lead him towards attainment of excellence in his skill.

## Adequate Provision in Curriculum

Science is a subject to be learnt by actually doing. There should be adequate provision in the curriculum for first hand experience through laboratory work or field-trips. A well-designed curriculum has a well-organised, well-graded, continuous, and practically attainable course content. The science curriculum must conform to the aims of teaching science, and must consider the age, stage as well as interest and abilities of the children for whom the curriculum is being framed. It should indicate the desirable behavioural changes expected of the learners as the final outcome of the curriculum. Each subsequent unit or item of knowledge must have link and relationship with the preceding unit or item of the contents.

The contents should be organised in order of difficulty. But this does not mean that acquisition of knowledge is the only purpose of the curriculum. As already explained, the process aspect of science must be given equal, if not more, weightage than the product aspect. Without the development of the values of learning science, science education will not only fail in its purpose, but may also become detrimental. A good and well-planned curriculum should develop in the learners the power of thinking. It must also provide for the study of the contributions and sacrifices of the scientists of the world.

The learners should be in a position to appreciate the sufferings of the great men of science while working for the benefit of the mankind. The curriculum must also provide opportunity for productive utilisation of leisure time of the learners and adequate locally relevant items of scientific activity to be pursued by the learners as hobbies. The curriculum of school science has long been criticised by different bodies and commissions like Mudaliar Commission (1952-53) as well as Kothari Commission (1964-66).

The Secondary Education Commission (Mudaliar Commission) criticised it for being narrow, bookish and theoretical, examination-oriented and unrelated to the child's environment. Remarking on the status of school curriculum in India in general, the Kothari Commission said: Against the background of the striking curricular developments that are taking place abroad, the school curriculum in India will be found to be very narrowly conceived and largely out of date. Education is a three-fold process of imparting knowledge, developing skills and inculcating proper interest, attitude and values. Our schools are mostly concerned with the first part of the process, the imparting of knowledge and carry out even this in an unsatisfactory way.

The curriculum places a premium on bookish knowledge, making inadequate provision for practical activities and is dominated by examinations, external and internal. Moreover, as the development of useful skills and inculcation of right kind of interest, attitudes and values are not given sufficient emphasis, the curriculum becomes not only out of step with modern knowledge but also out of tune with life of the people. There is thus urgent need to raise, upgrade and improve the school curriculum. Further, this Commission suggested introduction of the rural learner at secondary stage to the ideas and practices of scientific farming and the activities and skills related to it, while in the industrialised areas the curricula should have a bias towards the technical and industrial aspects of experimental science and its impact on industrialisation. But we are yet to go a long way in this direction.

There are various reasons for which we find the subject-centred curriculum still persisting and science lessons being read and memorised even today. The Education Commission made valuable suggestions for the improvement of social science curriculum.

The importance of the study of science for national development is well reflected in the National Policy on Education, Government of India (1968), wherein is specifically stated "... with a view to the national economy, science education and research should receive high priority. Science and mathematics should be an integral part of general education till the end of the school stage." Curriculum indicates the broad framework of learning experiences.

The details of activity programme for actual classroom implementation in terms of attainable units is called a syllabus. A syllabus, therefore, gives the details of the course as prescribed in the curriculum to be learnt at a particular stage. It is not possible to have a course content valid for all times. A syllabus, like curriculum, is also serviceable for a certain duration of time. It has to be constantly reviewed and reorganised. Since there are many variable conditions in the teaching process, it is impossible for any syllabus to provide for the needs of all the areas, all schools and all the pupils.

Every school has its own scheme of work depending on the need of the area, quality of the staff, situation and type of school and the material and human resources available. Different schools may like to follow different methods of teaching to suit the existing circumstances. The pupils, too, vary in their age, ability and aptitude. So also, the teachers in their ability and the success of teaching different topics depend on their likes and dislikes.

Moreover, it is one thing to have a course in detail for study and another to teach it successfully. However carefully a syllabus is made, it cannot guarantee successful teaching, for it is not possible to evolve an ideal syllabus to suit all circumstances and serve for all times. The selection and organisation of the science content is an important factor in determining the product of science learning. Usually, the teacher depends upon the textbook or the prescribed course of study for content-matter. Often, the textbook either contains extraneous matter or lacks in some vital up-to-date information. But even within the prescribed course content, the teacher will, to some degree, have to select and organise.

## Good Outcomes of Science Learning

The science content must not only serve the functional purpose but also inculcate the good outcomes of science learning. In the latter case, the content serves as a means to an end. The contents of facts, concepts and principles, in addition to being functional, should bring about behavioural changes in the learners. The syllabus, in addition to the topics of content, should include experiments, demonstrations, as well as outdoor activities to be pursued. The content for elementary stage should be selected:

- in terms of the broad concepts and principles of science;
- in terms of the interests and the needs of the children;
- taking into consideration the age level of the children and the level of difficulty for the children.

The contents selected should be related to the local environment with which the children are familiar.

The science course for children not yet mature should be elementary. The purpose at this stage should be to acquaint the children with the world around them. The science course should be confined to the study of nature such as plants, animals, common objects, simple natural phenomena, etc. It should also include an elementary knowledge of personal hygiene, physiology and elementary study of soil. The pupils of the middle stage undergo rapid growth and there appears an accelerated development of motor skill and ability.

Their interests widen and the period marks the development of keen desire to explore, discover and excel. They are more interested in sports, teamwork and collection. They are curious about adventure and mystery. They desire some freedom too, but always feel the need for security. While selecting contents for the middle level, the above facts have to be kept in mind.

The contents selected should:

- have use in their everyday life;
- lead to the inculcation of the desirable attitudes and outcomes;
- be appropriate to their level of ability;
- be in harmony with their needs and interests;
- encourage development of desirable social behaviour;
- contain matter from their familiar environment; and
- be continuous with the elementary science course.

At this level, elementary knowledge of physics and chemistry should be introduced.

Important related topics should be selected from the biological sciences (botany and zoology), physical sciences (physics and chemistry), physiology and hygiene and domestic science (for girls). This course should serve those leaving school after the middle stage. There should be a provision for practical (individual or in groups) work and other group activities. Though the subjects and the course wiil be similar to the elementary level course, the method of approach and teaching will be different.

More emphasis should be on activities. It should be remembered that the purpose and function of the middle level general science curriculum is to introduce the pupils to the broad fields of science. Thus, the general science curriculum for the middle level should be broad-based for familiarising the pupils with important fields of science. This is no stage for intensive study; neither it is a preliminary training for the high level science.

The purpose here is to give some insight to the subject of science as a whole. The adolescent stage is crucial. The understanding of the psychological developments in this stage is an important factor in planning appropriate curriculum for the High School level. This is also a period of rapid growth. Sexual maturity and a desire for independence predominate this period. At this stage, their special abilities and interests take a definite form; they experience a period of intense specialised interests.

## Syllabus Formation: Abilities and Interests

The syllabus for the high stage should, therefore, be framed on the basis of these abilities and interests. The Secondary Education Commission has suggested wide variety of diversified Courses for the pupils of high-stage (High and Higher Secondary schools). This, however, is meant for providing scope for development of special abilities and not for narrow specialisation.

Since only a small percentage of the total student population go to university or other institutions of higher learning, and the majority have to think of earning a living after the secondary stage, "the educational programme should give them some training in this direction, not so much in terms of the specific vocations as in training their practical aptitudes in preparation for definite vocational work" (Report of the Secondary Education Commission). Two types of science courses have been prescribed for the secondary grade pupils:

- a compulsory course of general science for all pupils imparted as a part of sound liberal education;
- courses of special science subjects, which the pupils may choose according to their aptitudes and interests.

The topics for general science were selected from the following branches of science to form an integrated course: astronomy, physics, chemistry, biology, geology, physiology and hygiene. The subjects prescribed for special science course were: physics, chemistry, biology, mathematics, geography, elements of physiology and hygiene (for those who do not offer biology). It has been mentioned earlier that the contents of the above subjects arise from the vocational needs and the future career of the pupils.

General science teaching has, however, a broader purpose in view. It forms a part of the general education of the pupils. While teaching general science, the procedure should be to introduce the subject through familiar things and pupils' own experiences. The illustrations should be from everyday life and local environment.

The topics or the units of study in general science should be organised around the areas of pupils' experience. A judicious amount of practical work should follow theoretical teaching. Discussions about the lives of selected scientists make science learning more interesting and effective. To live and survive in this modern scientific world, the knowledge and understanding of the basic aspects of science and a scientific outlook of mind is vitally necessary not only to the experts in science but also to the non-scientists and the ordinary men and women.

Everyone should have at least the basic knowledge of facts, principles, concepts and information of science required for everyday living in a scientific society and to appreciate what scientists are doing for the welfare of the mankind and how science has influenced our everyday life.

## General Education's Necessity in Science

The necessity of general education in science as a part of liberal education was conceived by Plato long ago. He said that the origin of both literature and science was the same and that though they seemed to diverge in technical methods, they converge in purpose. He advocated teaching of science not for its own sake but to develop an attitude of mind which could enable learners to apply scientific method to social problems. He thus conceived science as a means to an end. In society, learners who undergo school education need not become experts in science. Only a small percentage of students take up science as a profession, but all of them, whether they complete the school course or not, will have to live as a member of the society.

Moreover, the majority of the school-going children are of average ability. Hence, every individual must be given a general education in science so that he is able to live intelligently in this technological world. Since science has become a part of contemporary culture, we must teach science to all citizens as part of general education to provide the learners with common interests, knowledge and skill for effective participation in the developmental activities in our democratic society. The qualities and interests of the learners at the school stage are more or less common because of common geographical and social inheritance, common culture and common community life and a course of general science will benefit them all.

John Murray says that the general science is a course of scientific study and investigation which has its root in the common experiences of children and which includes the fundamentals of special science. He also says, "It seeks to elucidate the general principles observable in nature without emphasising the traditional division into specialised subject until such time as is warranted by the increasing complexity of the field of investigation by the developing unity of the separate parts of that field and by the intellectual progress of the pupils."

The course in general science has advantages over special courses. Here science is taught as a correlated subject not only with the various branches of science but also with humanities. It, however, does not mean that the students with special talent in science will be neglected. There is provision to cater to the needs of the specialists. The State needs experts and specialists in science and technology. Though all the students cannot be expected to become specialists nor it is the function of the schools to prepare specialists, the school cannot shirk the responsibility of encouraging and nourishing the budding scientists.

The school must motivate and help them learn and attain scientific abilities to the maximum of their potentialities. The Indian Education Commission has recommended disciplinary approach to science learning rather than the general science approach. The Commission felt disciplinary approach would be more effective in providing the necessary scientific base to the young learners. It suggested teaching of science as physics, chemistry, biology, geology and astronomy from class five of the middle level (higher primary stage). The Commission disapproved the general science approach to the teaching of science. The general science course mentioned above provides the basic grounding to the scientists as well as non-scientists. But the danger of over specialisation has not gone unnoticed.

The Ministry of Education, England in its pamphlet No. 38 states: There are dangers inherent in the situation; we risk becoming a well-fed rabble ordered about by experts; subjects of a new despotism. We cannot do without specialists; if we are not to become their slaves we must see to it that scientific knowledge and the ability to reason scientifically are widespread. This does not mean that all must aim at becoming technical experts; that would be impossible and in any event, few can become experts in more than a limited field. What it suggests is that science must find a place in liberal education, without aim of turning every one into specialists, in the same way as language and literature have done without any thought of turning every one into authors or critics. The content materials for general science should be:

- fundamental in nature;
- related to the local needs;
- related to the learners' environment;

- appropriate to the needs and interests of the learners;
- in harmony with the intellectual growth; and
- introduced through familiar experience.

After selection, the content materials are organised according to the correct procedure; because finally, it is the type of organisation that will decide the effectiveness of learning these materials. The forty-sixth Yearbook of the National Society for Study of Education (USA) gives the following guiding principles for organising the science contents:

- Contents should be organised into large areas or units, each representing a major problem of living, area of human experience or aspect of environment.
- The content of any single area or unit should be split into smaller learning problems which have interest, significance and usefulness to the learner.
- The learning experience in any single problem should be organised to promote functional understandings, growth in instructional skills, growth in the processes of problem-solving, and the development of attitudes, appreciations and interests.
- Abundant opportunities should be provided for applying and building concepts and principles.
- Provision should be made for effective evaluation including self-evaluation.
- The sequence of units should be planned to give recurrent contacts with facts, concepts and principles of science and provide a spiralling and enlarging pattern of growth in concepts and principles.
- Problem situation should provide definite training in one or more of the elements of scientific method.
- The course in science should be original and provide frequent opportunity for pupils to participate in planning and engage in individual and group projects.

The arrangement of the content matter may be made either in order of discovery, according to some logical order, or grouping a round some central theme such as a general principle in science or a local scientific phenomenon. But while doing so, there are other factors to be taken into account such as the teaching facilities in the school, type of the teaching staff, etc.

There are a wide variety of items in general science which can be arranged in either of the ways mentioned above. For example, a number of topics may be organised around the idea of 'gravity'; similarly, a large number of lessons can be established around a local 'hydroelectric project'. In the study of chemistry, lessons on 'element' should precede the lessons on 'compound'. This is quite logical. Any topic may be developed by following different stages through which it has undergone changes and developments since its early beginnings.

A study about 'atoms' can be started with the ancient idea of the Greeks, and followed with the idea of Dalton, Bohr and others up to the modern concept about the atom. This is an example of arrangement according to the historic order or the order of discovery. The UNESCO handbook on "Teaching of Science in Tropical Primary Schools" suggested organisation of the subject-matter of elementary science around the three broad areas—the earth and the universe and matter and energy.

The all-India seminar on the teaching of science in secondary schools suggested organisation of the contents of science under environment-centred topics, life-centred topics, and environment and life-centred topics, to bring about integration among the various branches of science. On the other hand the general science syllabus for the classes one to eight, published by the NCERT in 1963 developed the units of study around the following broad areas of contents (through major and minor concepts):

- Air, water and weather;
- rocks, soils and minerals;

- human body, health and hygiene;
- safety and first-aid;
- housing and clothing;
- energy and work;
- matter and materials;
- living things;
- plant life;
- animal life;
- scientists at work; and
- measurement.

The major and minor concepts are related to facts and events of everyday life and include necessary practical work as well as outdoor activities. The Board of Secondary Education, Assam, prepared the general science syllabus for classes five to ten, covering the entire secondary course in nine major concepts and 441 minor concepts. The Education Commission 1964-66 (Kothari Commission) has also examined in detail the different aspects of science education at the school as well as higher education stages. The Commission recommended that: In the lower primary classes, the focus should be on the child's environment, social, physical and biological. In classes I and II, the accent should be on cleanliness, formation of healthy habits and development of the power of observation. These should be emphasised again in classes III and IV, but the study should include personal hygiene and sanitation.

The child may also be introduced to formal areas of science such as the plants and animals in his surroundings, the air he breathes, the water he drinks, the weather that affects his daily life, the earth he lives on, the simple machines that are being used in his environment, the body of which he should take care and the heavenly bodies he looks on at night. School gardening is an activity that should be encouraged especially at this stage, as it provides pupils with direct and valuable experiences of natural phenomena. It is also recommended that in class IV, children should be taught the Roman alphabet. This is essential as the internationally accepted symbols for the units of scientific measurement and symbols for chemical elements and compounds are written in the Roman alphabet. Whatever one's language and the word for water, the chemical symbols for it is always $H_2O$. And it is far more than a symbol, it provides an insight about the nature of water.

Again, a knowledge of the Roman alphabet makes possible the use of maps, charts and statistical tables on an international scale. How expensive and time-consuming it will be to make available this material in all the local languages. At the higher primary stage, the emphasis may shift to the acquisition of knowledge together with the ability to think logically, to draw conclusions and to make decisions at a higher level. Science should now be taught as physics, chemistry, biology, geology, and astronomy.

The allocation of the above subjects among the three classes is suggested below; but other combinations may be tried depending upon the level of the students and local conditions: Class V—physics, geology, biology; Class VI—physics, biology, chemistry; Class VII—physics, biology, chemistry, astronomy. The general science approach to the teaching of science which has been widely adopted at the elementary stage during the last ten years has not proved successful as it tends to make science appear somewhat formless and without structure and runs counter to its methodology.

## Disciplinary Approach

A disciplinary approach to science learning would be more effective in providing the necessary scientific base to young people. The introduction of astronomy is specially commended, as it plays an important part in imparting good science education and in developing a rational outlook. From class V onwards the Indian almanac should be studied by observations of the night sky. Every primary school should have a science corner or a room to keep specimens, models and charts with the necessary storage facilities.

A minimum of one laboratory-cum-lecture room should be provided in every higher primary school. Science in the Secondary School: At the secondary level, science as a discipline of the mind and a preparation for higher education deserves special emphasis. In the lower secondary classes physics, chemistry, biology and earth sciences should be taught as compulsory subjects for all the pupils. Building on the introductory courses at the earlier stage, they should be made to cover wider areas and go deeper into the content than before.

The changing character of the sciences should be the major factor in curriculum development. Features of the Secondary School science curriculum: During the last few decades, the conceptual framework of physics has undergone a drastic change and this should be reflected in the high school physics curriculum. Similarly, in chemistry, the stress hitherto laid on memorisation of facts, formulae, processes and compounds should give place to an emphasis on the unifying concepts in the subject.

It is necessary to highlight the applications of chemistry in industry and daily life and its growing importance in our developing economy. Again, the present content of the school course in biology is traditional in nature. The concept of biology as a method of enquiry by means of accurate and comfirmable observations, quantitatively and mathematically analysed, and controlled experimentation should be impressed on the minds of the young learners.

Earth sciences should be introduced in the secondary school, geology and geography being taught as an integrated subject. There are also many areas in chemistry, physics and biology to which certain topics in the study of earth sciences can be naturally related. It has been pointed out earlier that the curriculum reconstruction is a dynamic process. Even a well-designed, well-thought out curriculum cannot continue to be applicable for all times.

This is specially true in case of science education. The growth and development of science in this modern world has been so rapid that it has become difficult for mankind to cope with the changes. The scientific knowledge doubles itself every decade. The concepts or facts of today may become out-of-date tomorrow. The curriculum, and hence the syllabus, must undergo constant research and review.

The curricula are becoming outmoded and inadequate more quickly in the recent times. This appears to be a world-wide phenomenon. As the Indian Education Commission stated that the school curriculum is in a state of flux all over the world today. Even in the most advanced country like the USA the reformed current curriculum is also being criticised and challenged. The recent trend in reforming curriculum development has been to shift emphasis of school science education from knowledge aspect to "interpretation and process" aspect.

The Physical Science Study Committee (PSSC) for physics, the Chemical Education National Study Committee (Chem Study) and Chemical Bond Approach Group (CBA), for chemistry and the Biological Sciences Study Committee (BSCS) for biology teaching were the study groups in the USA which during 1956 to 1960 studied curriculum improvement of secondary science programmes. Similarly, in the United Kingdom. Nuffield Foundation undertook projects for preparing improved curricula in physics, chemistry and biology.

In India, the Indian Education Commission of 1964-66 recommended a number of steps for improving science education in this country. In this connection, a special reference ought to be made to the remark of the Commission that while science must be pursued with full vigour to achieve Indian renaissance, the development of science must draw on the cultural and spiritual heritage of India. It disapproves following the western pattern, and suggests that the improvement programmes in science education should be based on the indigenous thinking reflecting "Indian ethos and value judgement."

**Basic Course**

The problems that arose for those charged with managing the curriculum at the school level can be analysed in three discrete categories: curriculum time allocation, teacher expertise, and resources in primary schools. The first reason for unmanageability is the apparently simple matter of time allocation. How much time is needed for each subject and for the whole curriculum? The problem has been commonly represented as the result of the statutory orders and the pint pot of the teaching week. If you are sitting on a national committee inventing or revising the statutory curriculum, it must appear merely a mechanical matter. Reduce the content until the whole curriculum matches, or is slightly less than, the time available.

The second reason concerns the task demands on class teachers attempting to deliver the whole curriculum-most obviously the range and level of subject knowledge required, the pedagogical skills necessary, including differentiation, the sheer detail and number of Statements of Attainment, and techniques necessary for reliable assessment. These have now been seen as unrealistic demands to make upon normal classroom teachers by all save the dismal succession of Ministers of State parading before us to assert that primary teaching, especially at Key Stage 1, is not a particularly demanding job, needing for its successful performance neither graduate knowledge nor long training.

The third reason concerns resources. Primary schools are staffed less favourably than secondary schools, even when the comparison is based on only Key Stage 3, yet the range and demand of their educational activities, and therefore their staffing needs, are almost identical. When Kelly ran a computer model of Stockport's curriculum led staffing, she found that it led to staffing needs that were more-or-less identical across the five to sixteen age range.

The first difficulty faced by teachers was simple; there was too much curriculum for the time available. One reason commonly given for this was that the statutory curriculum had been invented by means of committees of subject enthusiasts who were unable to consider how their subject had to fit into the whole curriculum.

There was some truth in at least the first two of these sentences, but the extent of policy error was substantially greater than Graham and Tytler allow. First, at an early stage Ministers had worked on a broad view that the Education Reform Act would create a curricular framework in which the national prescription should cover substantially less than the whole curriculum. This was because Religious Education and other important curricular areas such as health education and moral development were conceived of as lying beyond national prescription. The consultative document suggested that in schools where there was 'good practice' the National Curriculum subjects would occupy 70-80 per cent of curriculum time.

Three implications arose for curriculum management from the notional time allocations in this table. The most important is that they represented an error that was to have disastrous effects as the curriculum was implemented. The notional time for English and Mathematics combined, was dramatically lower than was conventionally provided by primary teachers, according to every study that had examined time allocations in the post-war period.

A summary covering much of the relevant research in the last fifteen years showed that 50 per cent of time was typically given to English and Mathematics, excluding their application to other subjects. Empirically, the phenomenon of 50 per cent of time on these two subjects—what we might call the 'basic instinct' in the primary curriculum—is firmly established.

Indeed, a study by Meyer *et al.*, which is an examination of official elementary and primary curricula world-wide across this century until the late 1980s, argues that the phenomenon has been a global constant-irrespective of region, political economy or state of development.

The second point refers to the time available for the whole curriculum. It relates to 'evaporated time', a term coined in a professional development programme and used by Campbell and Neill to indicate time technically available for teaching but used for non-cognitive purposes such as supervising children changing for PE, moving them from one location in the school to another, lining up and clearing away.

It is important because it is assumed to be available for teaching, but the studies referred to above that took account of this time by excluding it from teaching time, showed that something between 22 per cent of teaching time evaporates in this way, the amount depending upon the age of the pupils and the physical layout and size of the school and the methodology of time analysis. In general, younger pupils and those in open-plan settings experience more evaporated time. It occurs in small units of time in any one day, but in Campbell and Neill's research, based on time logs of over 3000 days from over 300 primary teachers, the average evaporated time per week was calculated at nearly two hours per week, equivalent to nearly 10 per cent of the teaching time available and equal to the notional time allocated for at least one of the non-basic subjects, in the National Curriculum.

Long hours spent on repetitive computation exercises do not necessarily mean challenging learning for bright pupils any more than do long hours spent by low-attaining pupils on tasks too difficult for them. Nonetheless, the conclusion from the research findings on time showed no general problem in primary classrooms about the adequacy of time being spent on teaching and learning the two basic subjects. Evidence after the introduction of the National Curriculum suggested that little had changed, with Campbell and Neill showing 51 per cent and 49 per cent of time devoted to the basics. The reasons for this state of affairs are three well-known ones.

First, as Ashton's research showed, drawing on a national sample of 1500 primary teachers in the early-1970s, the highest curricular priority was given to the basic skills of Reading, Oracy, Mathematics and Writing. Art. PE, Music, Sex Education, Science and Technology and a second language were given low priority. A follow-up study by Ashton with a less representative sample at the end of the 1970s, showed, if anything, higher priority placed on Mathematics and formal language competence.

Thus, commitment to the basics has always been at a premium in the professional culture. Second, parents and governors place highest curricular priority on the basics also, as Thomas's investigation of London schools showed, and the Government has required greater teacher accountability to parents and governors. Third, there has been a long tradition, stretching way back before 1988, of formal testing focused on Reading Comprehension and Mathematics, a tradition reinforced by National Curriculum assessment arrangements, though it is now slightly broadened by the inclusion of Science in the core. Such testing washes back into curriculum priorities.

The basic instinct is sustained and reinforced by the workplace culture in primary schools. Thus, it becomes startingly clear that the National Curriculum in primary schools, introduced by a series of Secretaries of State who had constantly banged on about the need to get back to basics and to raise standards of literacy and numeracy, put in place policy guidance designed to reduce the time typically spent by primary teachers on the basics, and especially on English, where the astonishing reduction of about one-third of existing time was proposed.

With respect to time allocations, the policy guidance outstripped the practice in the schools for liberality and breadth. It offered a fundamentally different concept and ideology of curriculum balance. The guidance given to the working group devising the English curriculum was such as to encourage infant teachers to reduce the time they typically spent teaching children to read and write from about seven hours to little over four hours a week. No wonder there was widespread stress at Key Stage 1 with infant teachers accurately reporting that, against their own professional judgement, they were having to reduce the time on hearing children read in order to fit in everything else.

As if to confirm this interpretation of the general policy, the National Curriculum Council issued guidance on planning the Key Stage 2 curriculum in which three presumably recommended case studies of school planning were provided: two of them (year 6 and 5 classes) suggested that the basics should be planned to occupy 41 per cent and 37.5 per cent of curriculum time respectively.

The Year 5 class plan was developed into a yearly programme in which Mathematics and English occupy 316 out of the 846 hours available, some 37 per cent of curriculum time in their terms. Schools are urged by the NCC to review the exercise, including reviewing whether the time allocations are appropriate. In this context, it is interesting, but confusing, to see that in the summer Update from the NCC, the curriculum planning in a Coventry primary school was celebrated; it initially used a plan which involved only 40 per cent of time on basics, but after training, it changed to 50 per cent.

There was thus built into the curriculum at national level a four-pronged problem for every primary school: too much planned prescription overall; too much content in the statutory orders for each subject: not enough flexibility to allow for evaporated time; a pretence or assumption that there would be more time available for the non-basics than there was in actuality. The task demands placed on class teachers by the National Curriculum have been analysed by Thomas, who shows that in the slimmed down curriculum, teachers at Key

Stage 2 had to be familiar with about 500 statements of attainment, detailed and confusingly presented programmes of study, poorly defined cross-curricular themes and religious education.

In addition they would have to possess subjects knowledge in the ten subjects up to about Level 6, or to be able to differentiate the planning, teaching and assessment across at least 4 levels or the curriculum in each subject. To achieve all this they would need to be the curricular school equivalent of Albert Einstein. Marie Curie and Lanford Christie rolled into one. The evidence about subject expertise sometimes takes a narrow view of the concept.

There are serious deficiencies in individual primary teachers' understandings, especially but not exclusively in the science and technology areas. Graduate students training for primary teaching provide no greater reasons for being sanguine, if those studied by Bennett and Carre are typical. They brought with them to their PGCE courses misunderstandings of everyday phenomena, such as the energy in a sledge moving down a hill, the explanations for night and day, and basic arithmetic competencies such as expressing $18 as a percentage of $120. Bennett and Carre's research used tests in Mathematics, English and Science based on National Curriculum levels 4, 5 and 6, and Assessment of Performance Unit (APU) tests. In some of these latter, the top 20 per cent of primary pupils scored higher than the average of the PGCE students.

Technically, of course, the curriculum demands are placed on the whole school rather than the individual teacher, but in practice for the medium term further, most teachers will remain responsible for teaching most of the curriculum for the pupils in their class. Given the research findings, Alexander, Rose and Woodhead's assertion that 'Teachers must possess the subject knowledge which the statutory orders require' sounds like a plea of desperation.

A major management issue therefore becomes how the specialist expertise in the whole staff group can be deployed to extend the work of class teachers, with the overarching purpose that, as Richards argued, such deployment should 'support not undermine' the class teacher's role. The previous two problems for manageability apply to all schools irrespective of location, size or other factors. Resource issues affect schools differentially, depending upon funding formulae, school size or pupil characteristics. There is however one key resource issue—the level of surfing.

The disadvantage for primary schools of historic staffing allocations has been acknowledged for almost a decade, although the difficulty of translating the acknowledgement into staffing resources has been enormous, particularly since the acknowledgement occurred in a period of public expenditure restraint.

If primary schools are to develop more specialist approaches to the curriculum, if class size should rarely exceed 30, and if teachers are to have some time free of class teaching in the school day; if, in short, the schools are to be enabled to implement the National Curriculum without stress and overload remaining a chronic feature of teachers' work, some improvement in the resourcing of schools is required. Acknowledgement is no longer enough. The funding formula at local level and the approval of schemes of devolved management at national level will need to be based on realistic assessment of the work activities now required of teachers in modern primary schools. Without the development of such national or local policies there will remain constraints on the extent to which the curriculum reform policy can succeed.

For the management of primary schools, the Dearing review's interim report appeared to hold out seductive promises to make the curriculum manageable by slimming it down. In terms of the analysis provided so far, however, it was limited to the issue of time allocations, and the evaluation of Dearing start form this recognition; the review had nothing to say about subject expertise or resourcing perhaps more fundamental problems than time allocation. Even the restricted problem-solving of the final report is a sleight of hand in respect of the 'curricular arithmetic' of time allocations, according to Alexander and Campbell.

Dearing's approach was to allocate 20 per cent of curriculum time for discretionary use by the school and to free two weeks of the school from any curricular prescription, and then to offer guidance on time allocation for the remaining time, viz for 80 per cent of thirty-six weeks. In discussing his own arithmetic, Dearing acknowledged that it was difficult to apply it to primary schools: 'Specific time will have to be set aside for work in English, but the full 30 per cent (KS1) and 25 per cent (KS2) does not have to be found in addition to time given to other subjects.' Perhaps too defensively Dearing added, "This is not sleight of hand; it is a statement of fact based on the realities of teaching expressed by teachers".

His own arithmetic was presented in a table in para, 4.20. With the bracketed figures referring to cross-curricular application, and therefore not counted in the total. Added percentages based on the total annual hours of 612 and 675 for which national prescription was envisaged, and these percentages appeared to reflect the historic practices of teachers, at least at KS1, with about 50 per cent given over to the basics.

There was however an assumed reduction in the proportion given to English at KS2. Alexander and Campbell however provide an alternative arithmetic, based on the overall time, not just the 80 per cent. Their arithmetic is given. Expressed this way, Dearing appears to have rendered almost a quarter of the school curriculum year discretionary, and therefore to have solved the manageability problem. However, as Alexander and Campbell show, there are five reasons why this has not really happened. First, the percentages for English and Mathematics are dramatically lower than previous or current practice, and time will have to be taken from the allegedly discretionary time to restore the basic instinct of 50 per cent time on them. Second, there is no educational or research-based justification for Dearing's assertion that as primary children become older they need less time on English.

The DES Primary Staffing Survey found 27 per cent allocated to English in the junior stage, and while it is obvious that the nature of English at KS1 is different from that at KS2 it is not clear why it should take less time. Nor has it done so in practice. Third, time on science has been reduced from what was previously proposed as necessary and the 10-15 per cent of time that was actually needed for the science curriculum. Fourth, national testing remains focused on the core subjects, pressing in on school management to ensure that teachers give priority to them in time allocation.

Alexander and Campbell calculated that all discretionary time would have to go to the core. Fifth, the time remaining for the other subjects will be inadequate, especially for those that are time consuming, such as PE, Music, Art and Technology. Sixth, without good reason, excessive time has been set aside for RE. As a consequence, the only way the curriculum will become manageable is if it is recognised for what it is, underneath the Dearing rhetoric a return to the elementary curriculum of the basic and RE updated with some Science and Information Technology. But manageability will have been gained at the expense of breadth and balance.

Management implications of the argument is based on two assumptions about the relationship between a national quango and individual schools. National quangos are not good at details, and have little competence in respect of the curriculum organisation of individual schools. For example, within the space of nine months the National Curriculum Council and the School Curriculum and Assessment Authority could not even agree on the simple technical matter of how many hours per year should be considered as available for instruction.

Second, as Thomas pointed out, the National Curriculum was established by reference to the best practice in each subject. To the Test or even good at everything is not common in individual human beings or their institutions so that there was and is a certainty that virtually and schools would, to some degree, fail to meet the statutory requirements. He added that 'Schools in other countries commonly fail to meet the requirements of their national curricula'. If these two assumptions are confirmed, the first lessons for school management are that it is in the individual school, not a national agency, that will create a curriculum that can be made to work for the school concerned; and that concern to cover the whole curriculum will have to be tempered by consideration of quality and standards in pupils learning. This latter goal might be more readily arrived at by the school establishing its curricular priorities than by attempting to deliver every statutory prescription equally well.

It was, after all, the quango charged with the inspection of schools that first made clear that coverage had been achieved only at the cost of depth in pupil learning. To restore confidence in their own professional judgement, it may first be necessary for teachers to lose faith in the ability of the national agencies to manage the curriculum implementation process sensibly. Dependency on external forces is as counter productive for the whole school staff as it is for the head and is likely to reduce rather than increase empowerment and accountability. On this basis six possible ways forward may be identified, though each will be contested by some teachers because of the value assumptions it contains.

First, in respect of time allocation, there is the possibility that much more of the teaching of English and Mathematics than currently might be planned, delivered and assessed through their application to other subjects. This is a position advocated for at least fifteen years by HMI, but has been found problematic in practice.

The problem is not helped by the framing of the curriculum in single subjects, nor by the tendency for mathematics and English schemes to be subject-specific. It is a good example of how a staff would have to have confidence as HMI had, that standards in the basics would improve by their being applied to other subjects. Unfortunately, it may be easier for staff in schools where pupils already have high achievement to develop such confidence than for staff where pupils appear to need substantial amounts of time on basic skills. Second, schools might consider the use of homework as part of their overall policy, especially perhaps at KS2. If there is shortage of time for English, part of the solution might be to cease to consider the pupils' curriculum time as synonymous with the timetabled school day.

Successful experience at KS1 with parental partnerships provides exemplars for developing similar approaches at KS2. Some opposition might be raised by those who would see such a systematic approach to homework as increasing the disadvantages of pupils who live in unsupportive families or physical conditions unconducive to doing homework. School based provision for facilities for doing homework, while a contradiction in terms, might go some way to mitigating the difficulties, though at a cost in staff supervision time.

A third solution to the problem of curriculum time might be to extend the school day, week or year. This is clearly sensible for those schools who currently spend less than the minimum expected weekly hours on instruction. It will be opposed on grounds of workload and of inappropriateness in rural areas where pupils have to travel long distances. It might also be surmised that while the current assessment and testing policy remains in place, any extra time would be devoted to the core and thus not release time for the non-core, which is the problem. A solution to teachers' lack of confidence and competence in subject expertise could be remedied by a greatly expanded Inservice training programme along lines similar to the DES twenty-day courses. This is obviously true, assuming that the courses are effective, and that their effects are long-lived, an assumption that Bennett and Summer's review of research questions in part. However, the problem for primary schools is that extensions of Inservice training programmes are also extensions of the times when teachers have to be away from school and their class, so that a substantial programme of Inservice Training in the school day may have disadvantages for pupil learning where schools do not have access to good quality supply cover.

A second possibility lies in suggestions that schools should deploy their staff in ways that enable their subject expertise to be exploited more effectively to the benefit of the school as a whole, whether as specialists, semi-specialists or as coordinators. Again, within limits, this might offer gains to some schools, especially those with staffing allocations that permit some flexibility of deployment. For most schools, however there is almost no flexibility, and, as school size reduces, the range of subjects in which there is staff expertise also reduces.

A third possibility is that more use should be made of class texts in which the teachers can have confidence that the intellectual content is reliable, freeing them to concentrate more upon planning for how pupils may learn more effectively from the texts. There may be some reluctance in the profession to the purchase of texts for whole classes, since it is seen as at odds with concepts of good practice which stress learning from first hand experience, building on pupils' interests, and practical investigation. While recognising that purchasing class texts will not solve all problems, perhaps influenced by past images of whole classes sitting reading class textbooks at the same pace, each pupil being asked to read aloud with the rest of the class following.

The way texts are used with primary classes is obviously an important matter, but the adoption of good quality class texts would be the quickest way of helping class teachers cope with both the cognitive demands of the whole curriculum and the time demands of preparing all learning materials for their pupils.

There is something puritanical in a professional culture which implies that the virtuous teacher is the one who prepares every worksheet herself. One of the few ways that a national quango could make itself useful to teachers would be to issue 'kitemarks' to texts to indicate that the statutory orders were accurately and adequately covered by them and that there were useful assessment activities integrated into them. This is, politically speaking, quite different from requiring that schools use one particular prescribed text.

Little use can be added about the formulae by which, external, resources are allocated to schools, beyond what has been argued earlier. On the internal allocation of the resources when allocated, one further point needs to be raised. In most primary schools lack of time, especially lack of time in the school day, has been a major obstacle to effective delivery of the curriculum.

Yet Campbell and Neill showed that teachers spent about eighteen hours a week teaching, and between five and six hours a week on low level routines, such as mounting displays, supervision, moving pupils round the school, registration, and collecting dinner money, etc. Where teacher had more time with a non-teaching assistant, they tended to spend more, not less time on such routines, probably because of the 'collaborative cultures' operating in them.

Yet there is something unsettling about the picture of teachers naming lack of time as the main obstacle to achieving the cognitive objectives of the curriculum while spending so much time on non-cognitive routines. Given that teacher time is the most valuable and most expensive resource available to a school, it is worth the management of the school exploring the advantages of rethinking the use of time of all adults on the staff of the school, to see whether the non-teaching assistants' time might be used imaginatively to free up teachers' time. Professional groups validate their professional claim by reference to 'expert' knowledge, supposedly hard-won and rigorously tested, but few people are naive enough to believe that in a job as complex as teaching such formal training provides the sole resource for subsequent professional thought and action.

Experience plays a substantial part. 'On the job', teachers develop skills and insights through constant interaction with children; they come to recognise patterns, commonalities and recurrences in behaviours, situations and problems and thereby develop habits of diagnosis and response whose practical effectiveness they can confidently demonstrate.

Similarly, to an extent perhaps not sufficiently recognised, teachers draw on their pre-training experience; for, uniquely among professionals, they are engaged in a process with which they have already been involved, continuously and unremittingly, since the age of five or earlier.

Their teacher training course is but a brief part of this total, cumulative educational experience and the latter, as much as the specifics of the one to four year training course, shapes the view of the educational process and its purposes within which they operate as teachers. It is a commonplace that many young teachers teach as they were taught in school rather than as they were urged to teach in college. Significantly too, teachers themselves display ambivalent attitudes towards initial training.

In the context of debate and negotiation about salary and status the teacher is a 'trained professional' in possession of expertise denied to all but those who submit themselves to the rigours of Certificate, B.Ed, or PGCE. But in everyday discourse, good teachers are 'born, not made' and that same expert knowledge may be dismissed as 'irrelevant' theory.

The truth is that it is empirically impossible to isolate initial training from earlier, contemporaneous or subsequent experience for the purposes of demonstrating its precise impact on the way a teacher performs in the classroom. It seems sensible to assume that it does have an impact, but to avoid the extremes of the grandiose strategic claim prepared for negotiation with Burnham on the one hand, and the dismissiveness of teacher folklore and Staffroom conventional wisdom on the other.

Instead, two hypotheses can be supported. One is that by incorporating in its emphasis on certain sorts of knowledge and skill particular views of the teaching role, the nature and needs of young children, the initial training course tends to facilitate some lines of subsequent professional development and to discourage others. Second, initial training influences subsequent development as much by what it omits or does badly as by what it treats positively; or, to use for the sake of convenience some mild jargaon, a course may 'de-skill' as well as 'skill'.

This latter point is particularly opposite in the present context, given that we have seen how primary ideology may relate to professional insecurity, The ideology, it will be recalled, is most strongly focused and most forcibly expressed in relation to those aspects of primary teaching where empirical study shows the greatest weakness.

Therefore, it is not firm evidence of behavioural casualty—element *X* in initial training produces action *Y* in the classroom—so much as a succession of positive and negative correspondences between training and subsequent practice which are sufficiently pervasive and exact as to leave little doubt about a causal relationship of some sort, albeit diluted by the power of experience, circumstance and contingency, and mediated through each individual's unique combination of personality, intellect and worldview. To provide a framework for the analysis which follows it is necessary at this point to give a brief resume of the overall structures and contexts of courses. There are two main routes into primary teaching, the four year B.Ed and the one year PGCE. The former is a post 'A' level undergraduate course and the latter, self-evidently, is taken by graduates in subjects which could equally well lead to other careers.

Deferred student choice and flexibility were major aspirations; this was the era of the 'container revolution', of courses which students put together from a wide selection of units and modules, and which often meant that professional study and work in schools were deferred until the second or even the third of four years. By the late 1970s, deferred choice, consecutive training and modularity were running into the logistical problems attendant upon contraction, for such courses had to be large to be viable. There were other criticisms chiefly revolving round the continued lack of sufficient professional emphasis, the perceived divorce of theory and practice and the split between subject and professional study.

A large number of institutions had transferred from university to CNAA validation and alternative notions of 'degree worthiness' had begun to be explored. Mark III B.Eds lasted the full four years, were usually honours only, and included serious attempts to put professional concerns at the centre of the course and at last to break down the barrier between subject and professional study. But by that time, during the early-1980s, there were countervailing pressures.

## Advanced Course

Successive HMI surveys, of primary and secondary schools had identified what were regarded as major weaknesses in serving teachers' professional expertise, above all in their curriculum knowledge. First HMI, then central government then ACSET, the teacher education advisory body-less from conviction than recognition of the irresistibility of political dogma backed by landslide election success-and finally the DES again proposed that all B.Ed students, whether primary or secondary, should spend half their course on main subject study in order to remedy these curricular deficiencies.

The inadequacy of the models of primary curriculum, primary teaching and initial training thereby encapsulated will be discussed. Courses are not disembodied artifacts but events and ideas which acquire their reality from particular institutional contexts. The majority of today's established primary teachers not only trained by the 2-3-4 year route but did so in a distinctive sort of institution, the college of education, which preserved a culture of remarkable homogeneity and historical persistence until the institutional reorganisations of the 1970s forced many of the surviving colleges into a usually reluctant alliance with mainstream higher education institutions.

While the PGCE was located mainly in the universities and until the rude advent of compulsory initial training, comprehensive schools and mixed ability teaching-prepared its students for grammar and public schools, the 3-4 year course reflected the requirements of a less prestigious tradition, that of secondary modern and primary schools. Thus, the two routes embodied and reinforced the mutual exclusiveness of the two central traditions in British education: minority/elitist/academic, and mass/elementary/utilitarian. Like primary schools, the colleges' origins were humble and impoverished. Like primary schools, they acquired a substantial contrary ideology-idealist, romantic, espousing values of self-actualisation, individualism and student/child. By the 1950s and 1960s, the professionally formative years for the deputy heads, heads and advisers of the 1970s and 1980s, the college ethos was predominantly one, in Taylor's often-quoted words, of 'social and literary romanticism'.

Partial rejection of pluralism: suspicion of the intellect and the intellectual; a lack of interest in political and structural change; a stress upon the intuitive and the intangible, upon spontaneity and creativity... a hunger for the satisfactions of interpersonal life within the community and the small group, and a flight from rationality. The extent to which this incorporates caricature is debatable, but when one considers the institutions into which most of the products of these colleges went-primary schools-the correspondence is irresistible.

Specific manifestations and echoes of these values will continue to emerge from our discussions. We turn now, however, from general background to the first of several specific aspects of initial training; the means whereby it seeks to generate that 'understanding of children' required for primary class teaching and pre-eminent in the class teacher's professional claim.

In most post-war teacher education courses the intending teacher's capacity to understand and relate to children has been seen as the virtually exclusive concern of two elements:

1. academic-courses in what until the 1960s were termed 'principles of education', subsequently the separate 'disciplines' of psychology and sociology of education, but more recently somewhat disguised within integrated, thematic education /professional courses;
2. experiential-teaching practice and other school based activity.

Little or no overlap of function is envisaged or, in terms of academic territoriality, allowed. There is assumed to be an exact correspondence between professional attribute and course component so that each becomes the 'property' of a particular department or group of staff. Thus, 'understanding of children' is the concern not of anyone with insights to offer but of just two academic disciplines, psychology and sociology.

Psychology is relatively well-established in initial teacher education. The McNair Report's 'principles of education' included; physiology and physical education, psychology, 'great classical writers on education', history of the education system, and appreciation of the 'home circumstances of the pupils'. This approach lasted well into the 1960s: Taylor, Tabble, Browne and others record the dominance of the 'mother hen'— the education tutor dispensing a mixture of 'method', history of educational ideas and, above all, psychology.

The 1960s witnessed the coming of age of the 'four disciplines' of education, but this strengthened the position in primary training, where the 'child development' course continued to rule supreme. Despite this relatively long-established pre-eminence in teacher education, it must be recalled that as an academic discipline psychology is young, and sociology younger.

Through a combination of empirical research and theory generation they seek to offer descriptions and explanations of individual and collective human behaviour which must be regarded as tentative, provisional and incomplete. Theories and models are frequently put forward, or at least received by students, as unassailable truths about the real world, their status as such apparently confirmed by the strong positivistic orientation of a good deal of the research drawn upon and by the convenient tendency of such work to offer quantified findings. Wilson showed how the very examination questions B.Ed and PGCE students were required to answer presumed their tacit acceptance of a wide range of concepts and constructs. That claim is readily substantiated.

Equally unsatisfactory, none of the questions invites application to the task of the teacher. Assuming the intending teachers duly demonstrate that they have committed the various theories, facts and arguments to memory, what then? What are they supposed to do with this knowledge? If it is seriously intended that it should inform their thinking about the job of teaching, why is no opportunity given for this capacity to be demonstrated? Or is it more important that they have knowledge than that they can use it? Taking their lead from this style of questioning, essays are peppered with the catch-phrase 'Research has proved that ....' without apparent regard for the need for all proof claims to be probed, for the provisionality of scientific findings. The connection with everyday professional practice is evident.

The same formulae re-emerge in much of the published work exemplified in, and in the written and spoken utterances of some serving teachers. Here, however, their linguistic hardness may be duly softened to match the gentler, familial ambience of child-centred discourse, and with the 'authority' now accorded a hushed, parent-surrogate, almost Messianic reverence: 'Piaget has shown us that...'. In both contexts such unconditional deference, by negating the element of natural scepticism combined with informed critique vital to academic study and the proper use of academic research, effectively invalidates the latter's claim: for it is no longer knowledge-open, provisional, challengeable dogma.

Conversely, the students are exhorted not to trust 'mere opinions', to have a higher regard for academic than common sense modes of analysis and explanation, and to prefer the 'objective' data of the social sciences to their or an experienced teacher's subjective' judgement. Understanding oneself and others has probably always been a human preoccupation. Certainly from the time when the first written record was produced we have shown a deep interest in human and animal behaviour. Yet our ideas have been almost entirely unsystematic and unrepresentative.

Even now we casually watch others or listen with prejudiced ears to conversation and from this evidence build up distorted rules of thumb about human nature. It is of course highly probable that our 'interest in human and animal behaviour' predates written records, but that is to quibble. More problematic, it seems to me, is the implied dismissal of all but the psychologist's way of doing things as 'unsystematic', 'unrepresentative', 'casual', 'prejudiced', and 'distorted': clearly not the author's intention, but open to that interpretation by someone new to the discipline.

Thus, may be generated or reinforced a basic epistemology to which the polarising of 'objective' and 'subjective', of 'fact' and 'value', of 'truth' and 'falsity', of 'knowledge' and 'belief', are fundamental. If internalised, this simplistic conceptual map is able and likely to provide signposts for a wide range of contexts: the teacher's subsequent response to educational research and theory most obviously, but also other situations in which knowledge claims are significant-record cards and diagnostic or attainment tests for example, and the wide and crucial range of claims which teachers make about children, their abilities, their potential, their home background, and so on. The other context where this epistemology bears fruit in a palpable way is the primary curriculum. Its most public and assertive face is the view of curriculum in general and knowledge in particular which we explored-for example the way a view of knowledge as brute 'fact' can be used as a justification for rejecting knowledge in any guise.

Less obviously, but perhaps in the end more significant, the framework may influence the way in which different curriculum experiences are presented for the child, and the view of knowledge the child thereby acquires: art at the 'soft'„ 'subjective', 'value' and the continuum science at the 'hard', 'objective', 'fact' end. Moreover, by according experientially derived insight lower status an initial training course misses an obvious and significant opportunity. Given that it is at the commonsense, intuitive level that the student/teacher is frequently forced to operate once under the pressure of everyday classroom circumstances, it is precisely these sort of judgements which should be exposed and explored during initial training, with a view to refining them and making them as reliable and reflexive as possible.

The psychologist's rejection of such perspectives is the more emphatic for being made on methodological grounds: personal knowledge is counted not so much less significant as inadmissible. Psychology has sought to replace commonsense theories about mental processes with propositions derived from the application to human nature of the methods of the natural sciences. Pre-eminent in this methodology is the charting of observable behaviours.

The 'introspective' method which seeks to uncover individuals' private knowledge, beliefs, attitudes, and so on by eliciting these by word of mouth, is considered by the behaviourist majority to be inconsistent with the scientific claim. Thus, because by the canons of a particular methodology such date is deemed inaccessible, as the object of study it ceases to be of interest. This raises broader issues concerning the historical development of psychology — its origins in the philosophy of mind, the late 19th century rejection of mind in favour of 'scientific' study of the brain and the central nervous system, the consequent issue of the distinctiveness of psychology vis-a-vis neurology and physiology, and the continuity of the behaviourist/ introspectionist debates. Such issues are beyond the scope of this book, but what is important for teachers and teacher educators is an awareness of the consequences for the way their task is defined in initial training. For teachers are not neurosurgeons: their main focus of concern is that elusive entity, which causes psychologists such difficulty, the human mind.

Alternative perspectives on children which would complement the portrayal of the behaviourist psychologist-from literature, drama, philosophy and pre-eminently, everyday discourse-are explicitly rejected. Apart from its tendency to impoverish the teacher's professional development and classroom thought, this monopoly reveals the extent to which the view of teaching as science has pervaded academic and professional opinion, even including groups-like literature teachers and tutors who might be expected to be more resistant.

Perhaps, as academics so often do, they fail to make the connection between the claim that the arts offer unique and profound insights into the human condition and the obvious fact that teaching itself is nothing if not concerned with that condition. Three provisos must, however, be expressed, lest it be thought that this section's discussion stems basically from an anti-psychology standpoint. The first is that the teacher education community as a whole, rather than its psychologists alone, have to take responsibility for excluding alternative sources of insight into children in general and into their mental processes in particular. There has been large-scale connivance at this needless impoverishment of the training process. Second, arguing for courses as a whole to include addition perspectives not for psychologists to do what artists do.

The third proviso is that the psychology component of teacher education courses, particularly until, in the mid-1970s, it began to be taught by graduates with a broad psychological training, may have been singularly unrepresentative of the parent discipline. It might, for example, over-emphasise Skinnerian behaviourism or developmental psychology; it might neglect study of the unconscious mind or of the social dimension of behaviour. It might fail to develop in students a proper consciousness of the extent to which a psychological model is a metaphor for behaviour, not the behaviour itself nor ever necessarily a particularly accurate representation of it.

Above all it might fail to convey the necessary sense of psychology, as of every discipline, as variegated, contentious and changing. The extent to which psychology of education courses directly reinforce the 'sequential developmentalism' element in primary ideology, which-drawing on King and others-provides a good example of several of the points above, particularly those concerning distortion in content and methodological oversimplification.

The central theme or core of such courses has traditionally been a chronological treatment of child development. Here 'development' is conceived as a matrix with norms for ages and stages providing one axis and various categories of human development —'intellectual', 'social', 'emotional', 'moral', 'physical', etc. This developmental matrix, open to fundamental criticism as it is provides a basic, widely accepted structure for primary discourse, curriculum planning and pupil assessment: the firmly fixed reference points on an otherwise shifting and undifferentiated map.

The ages attached to stages postulated by Piaget are approximate, and that the inevitability of the sequence and the state-independent processes and mechanisms of cognitive development-equilibration, assimilation, accommodation-are more significant than any inferred chronology. Yet it is notable that in student essays, as in professional discourse and teachers books and curriculum materials, the stages as such feature more prominently than the stage-independent theory, despite the fact that an understanding of the latter is essential to using the undoubted insights of Piagetian theory to promote or accelerate learning.

A Piagetian approach to HMI's concept of 'match', for example, would demand that children encounter learning tasks which are slightly, but not excessively, more complex than their present understanding, and that without this element of 'stretching' the disequilibration necessary for learning will not be produced. Where students or teachers perceived development in terms of states rather than processes they will tend to wait for learning to occur 'spontaneously' or 'naturally' rather than seek as teachers to advance it, on the grounds that the child has to be 'ready'.

Just as the dominance of developmentalism in everyday practice is matched by the dominance of the developmental 'matrix' in educational psychology courses, so the traditional fare of sociology of education courses corresponds strongly with another element in professional discourse, the family and home as the prime or even sole causes of the child's difficulties or failures at school.

Until the early to mid-1970s sociology of education courses were dominated by the issue of the effect of family and social class factors on the child's attainment at school. Early studies of streaming pointed the way to the possibility that the school itself might be a contributory factor in the under-achievement of working-class children, but research on class-related socialisation practices, parental attitudes, language, and so on, tended to swamp such relatively slender evidence. Only with the 'new' sociology of education, with its two-pronged, ideologically committed concern to explore first the cultural loading of the school and its curriculum in favour of certain groups of pupils and, second, the nature of everyday classroom life, did alternatives to the family/home deficit model present themselves with much credibility.

A substantial literature concerning the effect of teacher expectations on pupil performance, teacher constructs and typifications, classroom interaction and teaching styles, now permits a more balanced appraisal of the relative impact of family and school, parents and teachers, on the child's educational career. Some of this work was referred.

The shift, however, is recent, and everyday primary discourse, in as far as it is demarcated and to some extent controlled by senior members of the profession such as heads and advisers, still appears to display fairly unreserved affiliation to family/home theories; certainly the confident professional assertions about 'good' and 'poor' parents and homes are part of the essential fabric of both staff room discussion and pupil record cards. What will be worth monitoring is the extent to which the 'new' sociology of education produces a discernible shift in the way children's learning difficulties or lack of motivation are explained as students of the 1970s gain headships in the 1980s and 1990s and seek to influence their school's 'philosophies'.

If, however, the family background theory is such an indispensable element in professional ideology-in that, for the weaker teacher in particular, it is fundamental to the preservation of his self-esteem, we can anticipate little movement overall.

Theories of child development and educability, it will be apparent, not only have ideological potential but need to be simplified, and perhaps even distorted, to achieve that potential. Thus, Piagetian theory may be interpreted as confirming a doctrine of 'readiness' rather than as challenging the teacher's ingenuity to provide the child with appropriately structured and sequenced learning experiences. Similarly, the complexity and tentativeness of, say, Bernstein's work on language and social structure may be ignored in preference for gross polarisations of 'restricted' and 'elaborated' codes which may confirm a student's or teacher's existing cultural stereotypes and prejudices. Both theories exemplified can be invoked to justify low expectations of children on the 'grounds' of age, or of social class.

In fact, much of the theoretical material regarded as indispensable in initial training is intrinsically extremely elusive and difficult to understand, let alone to apply, and especially so for the 18-21 year old with neither professional experience of schools and children nor a background of introductory study in the social sciences. However, its misinterpretation could be reduced or offset if the typificatory process which it appears to reinforce were itself the object of scrutiny on initial training courses.

But it is a characteristic of mainstream education courses in initial training that, more recent sociology and social psychology perspective apart, they tend to devote little attention to the teacher as such. This is particularly true of psychology of education courses: an examination of student texts end course syllabuses will reveal that most deal not so much with the psychology of education as with the psychology of the child, and that child's education is treated only in so far as it can be conceived independently of the person who is its chief architect, the teacher.

The child emerges with an identity shaped by a combination of heredity and environment, having characteristics which are given and immutable. There is little or no psychological analysis of adults in general, or teachers in particular; nor c f the teacher's contribution to that classroom character of the child which serves as the basis for the teacher's appraisal of him or her. The child's actions are presented, if only by default, as independent of the teacher's; a conception of the child is encouraged which is somehow independent of the person, the teacher, who does the conceiving. These tendencies can be illustrated by comparing two recent psychology of education textbooks.

The model is not of the interaction of minds and personalities, still less of teaching as dependent on teacher qualities as well as child attributes, but of the operating theatre: the teacher, as complete person and competent professional, works on the child's mind with the detachment of the surgeon working on the anesthetised body of the patient. Skill is presumed; the sole knowledge required is of the mental anatomy of the child.

Fontana's psychology for Teachers stresses in its introduction that 'no child's behaviour can be fully understood unless we study also the behaviour of others-teachers, parents, school friends-towards him', and, subsequently argues that within the context of the school the teacher is the most important influence upon the child. Despite this promising beginning, and the author's reservations elsewhere about traditional psychology of education courses, twelve of the sixteen chapters are devoted to the pupil, only three to interaction and teacher-child relations, and just one to 'teacher personality and characteristics'.

The latter is a brief summary of research on the behavioural characteristics of 'effective' teachers, contextualised in an acknowledgement of some of the problems involved in defining teacher effectiveness: about a page each on the teacher's emotional security, attitude,, styles and classroom talk-all crucial issues but far too briefly dealt with.

In the economic climate of recent years, publishers have been increasingly reluctant to take on other than obviously marketable basic texts and course readers in education. The original or unusual is squeezed out and the second- or third-hand comes to rule supreme. Primary education has suffered particularly from the flood of edited 'readers', many of them drawing repeatedly on the same rather limited pool of 'safe' articles. Thus, though one must not overstate the case, the public teachable, examinable face of the educational process is increasingly defined by a combination of academics' personal interests and market forces: the research which, fortuitously, happens to be feasible, interesting, fundable or available, and what publishers and entrepreneurial editors or authors see as likely to sell.

But that is to disgrace. 'Understanding children' is an attribute not of the object of that understanding but of the teacher who claims it. The teacher will perceive a child in a particular way not only because of the sort of person that child is but because of the sort of person the teacher is.

In the primary school, we remind ourselves, the classteacher system ensures that a child is so perceived for educational purposes for a whole year by just one person: with that much at stake it seems indefensible for initial training courses to neglect the psychology of the teacher. Four shifts are indicated therefore. First, the inclusion, as argued, of a substantial focus on teachers and their impact on those various aspects of the learner conventionally treated as independent of them.

Second, a deeper exploration of the ways individual behaviour-whether the child's or the teacher's, can be understood in the context of, and sometimes explained as a consequence of, interaction, possibly through the use of transactional analysis techniques applies in industrial psychology and psychotherapy, as well as interaction analysis schedules and theoretical perspectives of social psychology and interactionist/phenomenological sociology.

Third, a preparedness to explore the irrationality which frequently characterises human actions and interactions, not least in the classroom; the dominant psychological tradition in teacher education, as we have seen, takes the teacher's total rationality for granted and moreover imposes a sometimes over-tidy, predictive framework- on the child. Fourth, just as theoretical study of children in teacher education is required to be supported by work with children in classrooms, so teachers in classrooms and above all students themselves would need to be the object of practical study.

If self-exploration is now included in the training of other professional groups whose job involves the management of people and relationships, it can surely be justified in the training of teachers. The second major context a course provides for developing the student's understanding of children is school experience. Again, a historical perspective on this part of the course is helpful because its character and purposes have changed in recent years, though not as radically as some current advocates of 'school-based' courses would have us believe.

To make use of schools not merely as a context for practising executive skills but as a prime means for developing the student's capacities to observe, understand and relate to children is a *sine qua non* of teacher education. It has to be asked, however, whether the opportunities are fully exploited. This is only partly, as we have seen, a procedural matter; what needs closer attention is the teacher educators' and course validators' view of what this 'understanding' might mean.

At present, and certainly during the decade when today's primary teachers and heads trained, the interpretations, explanations and hypotheses of students and serving teachers have been undervalued and therefore insufficiently pursued and tested. And while a somewhat restricted canon of child related theories has been, in comparison, overexposed, it has been the exposure of obeisance, rather than critique.

Some of the consequences, or correspondences, suggested by this restricted epistemology have been outlined in this section and exemplified more fully. There seem to be a number of ways the situation can be improved, of varying degrees of radicalism:

1. the focus for 'understanding' can be broadened;
2. better use can be made of existing education disciplines;
3. additional (academic) sources of insight can be explored;
4. 'everyday' modes of understanding can be more fully exploited;
5. different conceptions of professional theory for teaching can be applied.

In an obvious sense, by definition, students are the main concern in a course of initial training: the qualities, skills and knowledge which they are deemed to need are the course's *raison d'etre.*

So while arguing that the course needs to focus more explicitly on the student and on the serving teacher, we are arguing for a specific kind of attention with which this self-evident concern is not to be confused. Courses are premised on the importance of teachers' mastery of certain executive skills, their manifestation of particular personal qualities and their possession of certain kinds of knowledge.

Pre-eminent among the latter is that academic knowledge about children which we have explored in this section, and which they may or may not make significant use of in practice. What, by and large, are neglected are the many layers or facets beyond these generalised propositions which combine to create the particular ways individual students and teachers actually view, or 'understand', the children they teach in particular classroom settings: teachers' actual and tacit, as opposed to preferred or espoused or idealised, knowledge of children, and the biographies which produce this knowledge: their subjective realities, as opposed to the quasi-objective ideas or the educational situation with which they are presented in training.

Merely to offer to the student a set of propositions from psychological/ sociological theory or research-this is the way children are, this is what they are like, this is the reality to which your decisions must be addressed is to ignore, or at least to fail to acknowledge sufficiently, two basic arguments concerning children in classrooms in which much of this book's discussion has been grounded. First, regardless of generalised principles of child development, motivation and so on, children are 'as they are' in classrooms, in part at least, because of the actions of the teacher; they respond, as in interaction all humans respond, to personality, to climate, to tacit or explicit signals, attitudes and expectations from a variety of sources, but chiefly, in the educational context, from the teacher.

Second, though individual children are viewed differently by each of those with whom they interact, and their self-concept indeed in part evolves from a consciousness of these various perceptions, in the educational context the teacher's view of them is the most significant. It is the teacher who defines their abilities, their potential, their personality, their attainment and their attitudes, for the purposes of making curricular decisions, evaluations and predictions, and, as we have seen, in a way which is consistent with their ideology. Such is the nature of the educational process, and the classroom power relationship, especially where young children are concerned, that the central assumption in child-centredness, of the child's autonomy, is not only fallacious but dangerously so.

It can never be the case that on the one hand we have the child, about whom there is pre-existing objective knowledge, and on the other the teacher, who simply has to acquire that knowledge. The teacher's knowledge of the child is subjective, it is created by the teacher, its character is therefore as strongly conditional on the way the teacher is as the way the child is. The theme for initial training, which should complement 'understanding children' is 'understanding how teachers "understand" children'.

And if it is indeed students understanding that we wish to promote rather than their capacity merely to parrot the prepositions and formulae of others, the source as well as the character of such 'understanding', on the basis of which students and teachers act, needs to be explored; this necessitates attention to the unique individual biographies of each students as well as to the more generalised analysis of the professional, historical and ideological situation of the particular groups of teachers offered. The outcome of this process should be to sensitise the student to the possibility of a much wider range of diagnoses and explanations. The fact that introspective methods are unacceptable to some psychologists is worthy of debate in this context but is not a ground for rejecting the perspective defined.

The course aims to train teachers, not professional psychologists, and if a perspective is helpful it should be included, methodological qualms notwithstanding. Thus, the scope of 'understanding children' in initial training needs to be broadened chiefly by taking in teachers' influence on both the children and their 'understanding' of them.

Equally important, and more easily enunciated, is the need to allow existing perspectives on children offered by psychology and sociology to be supplemented by alternatives from within those disciplines. Clearly, given our discussion of the complementary relationship between ideology, professional theory and practical situation, the introduction of less comfortable psychologies and sociologies might be resisted, but perhaps we can begin to accept that our comfort may well be secured at the expense of the quality of the child's education.

This shift has already begun in the sociology of education: the sociology of knowledge and of classrooms has provided the needed counterbalance to the family/home educability preoccupations of the 1950s and 1960s. We now need a 'new' psychology of education which more fairly represents the richness of mainstream psychology.

The newer perspectives offers both alternative focuses and tools for analysis. Centre-stage are not only classroom processes and the interactions of teachers and children which feature in 'objective' study in the Flanders tradition, but also the meanings which the teachers and children themselves assign to those processes. The underlying assumption here is that human actions can be properly understood only if one uncovers these meanings, since 'action is forged by the actor out of what he perceives, interprets and judges... The "objective" approach holds the danger of the observer substituting his view of the field of action for the view held by the actor.

In studying classrooms it becomes important to understand pupils' as well as teachers' definitions of the situation, and the extent to which the latter are not autonomous but are influenced by the former; at the same time the power differential in classrooms makes the impact of teachers' meanings on children and classroom life considerable. This perspective prominently informs the classic and comprehensive exploration of school interaction by Hargreaves and specific studies of primary classrooms like that of Berlak *et al.*

The study of infants' classrooms by King, to which I have made frequent reference, is eclectic: it is critical of phenomenological, interactionist and Marxist perspectives yet is also influenced by them. It belongs to the 'family' in so far as it is grounded in close and sustained observation which is interpreted by reference to actors' perceptions and explanations rather than observer preconceptions.

At the same time it places these in broader frameworks of ideology and social structure. These developments are recent: empirical study of primary classrooms, of whatever methodological complexion, is still relatively thin. Nevertheless, in terms of what by this book's analysis seems to be needed—far greater attention to how teachers 'understood' children and teaching -the growth points are now significant, diverse and rich.

The last point notwithstanding, we also need to ask whether the monopoly by psychology and sociology of insights into children and classrooms is to be desired or supported. The other two disciplines in the educational studies pantheon are philosophy and history. Philosophy of education-whether so defined or used thematically in an integrated course tends to concern itself with broad non-contextualised questions about aims, the nature of knowledge, the justification for particular educational concepts and activities, the ethics of reward and punishment.

Frequently it is conceived as a tool of encouraging a sharper, more reflexive and considered mode of analysis than the easy, instant judgement, as a basis for critique of teachers' and students' 'commonsense' statements. This is undoubtedly necessary, but it is also the case that by such means, intentionally or unwittingly, is the status of academic thought preserved, for philosophical analysis ought equally to be applied to the statements, judgement and explanations offered by psychology and sociology of education: conceptual analysis should have no boundaries.

Especially, a concern with epistemology ought not to start and finish with the school curriculum, but should encompass the knowledge and truth claims of the teacher training curriculum as well, the ways of making sense of and understanding children, teaching and learning which the initial training course expects the student to internalise and subsequently 'apply' in the classroom. In other words, 'ways of knowing' about educational processes should be subjected to the same level of scrutiny as are the ways of knowing which constitute the school curriculum out there in the school. To fail to do this is to miss an opportunity to give the student a working understanding of epistemological issues; and it could be construed as hypothetical to subject to critique the school curriculum but not that of teacher education. A similar argument applies in the case of history of education.

We saw that how usefully a historical awareness of the institutional and ideational background of primary education both illuminates and provides a basis for critique of present-day ideas and practices, particularly in respect of the ways of the child is viewed and the curriculum defined in the context of the class-teacher system. We also saw how primary professional discourse with its 'cocoon' imaginary and polarising of the child and society, seems resistant to a sense of the interplay of historical events, cultural values and educational ideals.

Conventionally, history of education courses have done little to remedy this. Courses in 'the educational system of

England and Wales' can still be as normative, functionalist, systemic and superficial as they were in the years following the McNair Report: presumes of the clauses of education acts and the recommendations of major reports, but rarely interpretations or explanation beyond a sort of sub 'O' level 'seven causes of the Boer War' variety, and certainly little real delving into the pedagogy and curriculum experienced by previous generations of children and the justifications offered by teachers and others in support of these.

Of course academic monopolies are not fortuitous: the one under discussion reflects aspirations to make teaching a 'science' grounded in a set of empirically derived principles and so to demonstrate the 'expert knowledge' basis of the teacher's professional claim. However, we might try asking afresh, with no preconditions, the open question: 'What is the best way to develop the young adult's capacity to understand other human beings, especially pre-adolescent children?' in pursuit of an answer we might attempt to catalogue those ways which humans have learned to understand each other and themselves. On the one hand, and prominently, there is the pervasiveness and potential of insight grounded in individual and collective experience.

Meanwhile, we concentrate more on academic or public modes of personal and interpersonal exploration, we have to acknowledge that the field can encompass, at least, literature, art, music, drama and religion as well as the social sciences; and that even the latter can extend much further than teacher education has allowed. The modes of enquiry we actually make available to the student are from one small spectrum of human knowledge, and, as it happens, they are from one of the newest and-to workers in the physical sciences and the arts alike one of the most suspect in terms of its claims to represent humans as they are. Putting into operation such extended concepts of 'understanding children and teachers' is not necessarily easy, but the issue needs to be faced. As presently conceived, teacher education courses place arbitrary and unnecessary restrictions on this aspect of the student's development.

The most basic reappraisal would concern the academic/ professional distinction, and, as a consequence, the common ground between subject studies and education theory as regards each's capacity both to generate professional insight and to meet 'personally educative' functions might be disconcerting. However, the most substantial and necessary shift in this context is towards the exploration and use in initial training of non-academic, everyday, subjective professional knowledge.

The arguments seem inescapable: such knowledge is pervasive, inevitable and influential in everyday practice and therefore requires exploration; it is effective in that it is the basis for teaching of a high quality and therefore may encapsulate ideas worthy of emulation; it is also, conversely, the basis for weak teaching and therefore its limitations as well as its strengths need to be exposed. Whereas in this context, the everyday knowledge in question concerns children, it taps, or may tap, insights stemming from one of the most fundamental of human relationships: it cannot simply be disregarded on the grounds of arbitrary stimulative definitions of what constitutes 'science' and 'objectivity'. This is not an argument for rejecting academic enquiry, or for a revival of apprenticeship approaches to teacher education.

Rather, the case is made for recognising the strengths and limitations of any mode of understanding pursued in overmuch isolation-particular disciplines, academic study in general, personal, experientially grounded everyday knowledge. This includes acknowledging the advantage of using these in combination and juxtaposition, particularly in pursuit of that understanding of ourselves and others which educationists seem happily prepared to lay claim to, despite the fact that the rest of humanity has found it rather more elusive. Such eclecticism as is argued here would be conditional upon courses sensitising students to the epistemological problems raised. It is not adequate to 'raid the disciplines', or rather to 'raid' all available and potentially productive sources of insight, without also understanding the nature of the truth claim each makes and the limitations thereof.

Traditionally courses have tended to treat academic sources as given the experiential sources as suspect or unacceptable: all are problematic, though in different ways. It will be apparent by now that, in combination, the ideas above require not minor adjustment to the content and pedagogy of initial training courses, but a more fundamental shift.

'School-based curriculum development' is a clumsy term, used to refer to activities which, because they take place in the unique contexts of individual schools, are necessarily diverse. This section attempts to place some structure on the idea of school-based curriculum development by linking material from four relevant areas. First, theoretical concepts are outlined; second, these concepts are qualified by reference to the practice of curriculum development in primary schools; third, contextual factors influencing the nature of such development are examined; and finally, the values underlying school-based curriculum development in primary schools are briefly elaborated.

One of the problems faced by anyone attempting to understand school-based curriculum development as an idea is that it is used very much as a catch-all concept. For example, as illustrated by Miston, it means something as substantial as groups of secondary school pupils, together with appropriate staff training and resources development; or it may be small-scale revision of a language programme in a primary school, such that is described by Timms and Lees.

The DES have stressed the significance of school-based approach, but two writers in particular have helped to sharpen the formulation of ideas about them. Eggleston's introduction to six case studies provides a useful starting-point. Although it is unclear whether Eggleston was describing school-based curriculum development or prescribing a particular way of doing it, four features of his definition are worth picking out:

1. *It is particularistic:* The curriculum-development activity is focused upon the diagnosed, or perceived, needs of the specific school or part of it.
2. *It is process-oriented:* In the terms of 'strategies for the curriculum' intended, the process by which these are developed in important in itself.
3. *It is participatory:* The appropriate style for developing the curriculum is cooperative, that is, staff working together to produce plans for change.
4. *It is preliminary:* The curriculum developed is to be seen as experimental, in the sense that it is open to evaluation and appraisal after its implementation.

One of the interesting things about this definition is the stress, not on the curriculum as such, but on the roles that teachers have to play in the process of its development, and the attitudes that are required to underpin it. This characteristic is also central to the analysis offered by another writer, Skilbeck. He identified three models of school-based curriculum development the rational-deductive, the rational-interactive and the intuitive-and located them within differing politico-educational frameworks.

The first operates in centrally directed educational systems, where the task of the school is to 'interpret central directives' and the role of the teacher is as a mere functionary in a bureaucratised educational service. The second emerges in mixed systems, such as those in England and Wales, which stress the active role of teachers in adapting the curriculum at school level within rather broad general outlines of national policy. Teachers working within this kind of framework have a more complex role than in the rational-deductive model and have more demands made upon them; they 'have to act as course assessors, to help construct syllabuses, to select learning materials and to devise learning systems'.

The third model stresses the individual teacher's decision-making and creativity, and leads to great diversity between teachers, between schools, and to inconsistency between 'normal policy and individual school programmes'. Although school-based curriculum development may share elements from all three, the rational-interactive model represents the style and values most appropriate to contemporary English schools, especially because of the stress put upon the range of roles expected of teachers, and the assumption that teachers have to negotiate the fine details of their roles by working in partnership with each other.

Thus, in this formulation also, school-based curriculum development is as much about changing roles and relationships among a school staff as it is about changing schemes of work or methods of teaching and learning. A major goal of school-based curriculum development in Skilbeck's terms is the continuous adaptation by teachers of externally defined curricula into forms of educative experiences unique to the teacher and learner. We need a system for curriculum development that combines the advantages of national policy-making, national centres for the production of materials and for research and development, with the flexibility, adaptability, and professionally satisfying features of local initiatives and creativity.

Thus, school-based curriculum development is predicted upon the concept, admittedly idealised, of teachers who creatively reconstruct the curriculum within a recognised framework of local and national expectations; it is not predicted upon passive acceptance of external definitions of the curriculum, or the myth of the 'autonomous' school, existing independently of its political and economic context.

Given the political development, outlined, of a national framework for the curriculum, the relevance of Skilbeck's model for teachers in England and Wales has increased considerably.

Skilbeck's analysis enables us to add two further characteristics of school-based curriculum development to the four identified earlier. These concern the relationship between the school's curriculum and national or local guidelines, and the view taken of the role of the teacher.

The distinctive arena for school-based curriculum development is the staffroom rather than the classroom, and the distinctive discourse is concerned not only with the surface details of curriculum practice, but also with the assumptions underlying it. The theoretical analyses discussed above are necessarily generalised, and need to be qualified somewhat in the light of the practice in the Warwick enquiry schools.

Brief outlines of the programmes have already been provided, which suggest two characteristics that are perhaps distinctive to primary school curriculum development. They can be considered under two headings, namely gradualism and specialism. To dichotomise rather too simply, gradualism qualifies the idea of 'development', while specialism qualifies the idea of 'curriculum'. By gradualism, three related features of the notion of 'development' which stress the limited expectations that may be held for it. These three limitations may be thought of as the problematic, the unpredictable and the incremental qualities of curriculum development in primary schools.

The problematic nature of curriculum development derives from the fact that there is a conceptual difference between development and change. The latter is neutral and implies merely that practice has altered, not that it has been improved. 'Development', 'renewal' and probably 'innovation' imply not merely change, but change for the better. What counts as a change for the better in education is problematic.

An analogy with the practice of medicine may be helpful. If a doctor diagnoses, say, constipation in a patient and prescribes a change of diet as a remedy, there is not much professional or lay disagreement about what would constitute a change for the better in the patient's condition. Not being constipated is generally regarded as an improvement on being so. But the school curriculum is a more difficult area for diagnosis, with less sure a basis for agreement about what constitutes improvement.

A curriculum where there has been, so to speak, little movement for a number of years is not necessarily in a worse condition than one in which there has been a great deal of it. One of the points that follows from this is that, although the term 'curriculum development' is widely used throughout the literature and in this book, in practice it cannot be shown in advance, and often it is not known even in retrospect, whether changes are actually developments. At best it is commonly a matter of belief, intuition and professional judgement of those involve.

Indeed, to return to Eggleston's emphasis on the process of development, it was quite striking to note how frequently the teachers in the Warwick enquiry reported that the major benefit, for them, of school-based development activities had been the experience of being involved in the process rather than, or in addition to, any changes in actual curriculum practice. It was as though they were hedging their bets on the curricular outcome of the initiatives. Second, there is lack of predictability in outcome.

Given the experimental approach characterised by Eggleston as 'discussion, planning, trial and evaluation', there is the built-in risk that the evaluation might show no tangible development in the desired direction. This seems to have been the case in the Oxfordshire primary school reported in a case study for the Open University prepared by Cliff. He reported a self-evaluation exercise by the school staff, lasting over a year, and involving at least seven in-school staff meetings, to prepare a review of the school's policy, provision and practices across a range of activities, including the curriculum. It appears to have been characterised by serious, professional involvement by the staff, and to have been efficiently and yet flexibly organised. It covered, among many other matters, problems or curricular aims and how they should be described, curricular practice, including the grouping of pupils and catering for individual differences, and the provision of curricular guidelines, and was carried out by effective teacher collaboration.

The kind of sweeping changes in curricular aims and practices currently being promoted for the 14-19 years olds, for example, could not easily be developed in the school-based mode precisely because they are predicted upon external intervention fundamentally to redefine curricular aims at least in the innovation period. School-based curriculum development is a more modestly conceived activity, designed to build upon existing mainstream curricular practice, and predicated upon the assumption of a curricular framework about which there is already consensus, or which can be taken for granted. It is incremental, not radical, change in the curriculum, with slow, small-scale, almost routine, benefits, accruing over time from a school staff gradually building upon its collective strengths and, where possible, remedying weaknesses.

There should be little of the heady rhetoric associated with large-scale national projects of the 1960s and early-1970s, not just because such rhetoric tends to lack credibility in the routine of school life, but because school-based approaches are designed to improve the normal curriculum, not graft abnormal practices on to it. Discussions with teachers in the Warwick enquiry illustrated the incremental quality in an interesting way, showing the mundane and highly pragmatic nature of what was involved.

In School 1, a staff group met to review their curricular policy in social studies, and attempted to revise an existing scheme by constructing it around some basic concepts and skills, drawn partly from some Schools Council curriculum materials and partly from their experience with the previous scheme. No grand claims were made, or were felt necessary to be made, about either the changes that might follow or the process itself. Many frameworks were tried in private and rejected; it was felt important to demonstrate that the content could be organised-conceptually.

To test this we developed a matrix, setting four broad conceptual areas of Environmental Studies across four conceptual themes from 'Home and Family'. When we did this, we simply jotted down in the matrix what we thought would work, given that had worked before, and the overall conceptual scheme we were developing. And then we met and looked at what we'd written down and tried to sort it out from there.

What strikes one about these statements is their very ordinates, their tentativeness and the absence of extravagance is what is being claimed. They are the voices of routine improvement not radical, or even substantial, change, expressing the gradual, incremental quality of school-based curriculum development at primary-school level. They embody the view of HMI that: A slow but steady build up from the points of strength of individual teachers is probably the only way forward.

Specialism refers to the exploitation of expertise in a subject or, more accurately, in a curriculum area. This exploitation took a variety of forms in practice but three can be distinguished: specialist teaching, subject teaching and subject diffusion. Specialist teaching occurred in two cases, with teachers used to teach a number of classes and having no class responsibility themselves.

The two cases provide ambivalent evidence, limited though it is, for the idea of specialist teaching as a basis for in-school development. Both teachers taught their subject to all the older children in the school, but none the less needed to influence the quality of work with the younger children, who were taught by class teachers. They exercised responsibility for raising the quality of work done throughout the school, though they also acknowledged a reduced impact upon the younger classes. They appear to have experienced little of the conflict and strain that other postholders felt, as reported. This was probably because teaching one subject and not having class responsibilities dramatically reduced the range of the other demands made upon specialists.

On the other hand, as suggests, the two teachers used as specialists tended to have involved their colleagues less in the process of curriculum development than did other postholders. Furthermore, the specialist teachers did not involve colleagues in evaluation of their programmes. This suggests, accepting that these are two cases only, that specialist teaching, whatever its merits in classroom practice of the specialist, may, in terms of initiatives of a school-wide kind, tend to produce leadership of a more isolated, less collective style than would otherwise be the case. Subject teaching occurred when teachers with specialist subject knowledge taught one or two other classes on a regular basis, while retaining a generalist class-teaching role with their own class. This occurred in two cases.

A related version of subject teaching was when a teacher irregularly, and for limited specific purposes, taught alongside colleagues, or swapped classes, in order to teach a specialist topic or skill or to demonstrate a skill and show the quality of work that could be expected. This happened in three other cases. Subject diffusion occurred in all the programmes, when teachers with expertise were consulted by others who needed advice, or when they took a lead in curriculum planning and review groups.

It was the main mechanism for spreading specialist knowledge from the postholder to the other staff. It was what the Inspectorate meant when they talked of postholders' having an 'influence' on the work throughout a primary school. With organisational arrangements reflecting the class-teacher principle, even where some subject teaching occurred, this diffusion model was the dominant style of in-school development.

The stress on the use of teacher specialism in the curriculum development programmes is largely explicable in terms of the political analysis. All the programmes were focused on issues raised in the Primary Survey, which stressed the need to provide for greater progression and continuity, to cater for able children, to fill the gap in science, to extend advanced reading skills, and so on. Although it may not be a permanent characteristic of primary school curriculum development, it probably provides quite strong evidence of the impact of the survey, an impact followed up in two later surveys by HMI in which 'specialisation' and its contribution to curriculum development in first and middle schools were further examined.

There is an ecology of curriculum development. Just as certain kinds of plant and animal life flourish in favourable conditions of soil, light, temperature and the balance of relationships in their overall environment, so the context of primary schools inhibits or encourages the growth of the kind of curriculum development discussed above. Factors in this context can be considered as external to the school or internal to it, although the distinction is not clear cut.

External factors are those derived from the educational system itself, from its administration, finance, demography and from teacher career opportunities. Internal factors include the role relationships of teachers and the ways in which authority is exercised by the headteacher.

More important, however, overall staff expertise may be affected, because the impact of falling rolls will lead to loss of flexibility in the ways that staff can be used, and sometimes to arbitrary loss of specialism. Even if staffing is reduced in line with pupil numbers, a school's ability to deploy teachers in ways that can effectively exploit their skills and expertise will become limited and, most crucially perhaps, the opportunities to free teachers for necessary curriculum development activities in school time will be lost. It is for this reason that some local authorities have developed what they refer to as 'curriculum-led' staffing policies.

Although there is a suspicion that such policies could be used primarily to decide that small schools are not viable, an interesting consequence of them is that the local authority has to make clear what its conception of an adequate primary school curriculum is, for curriculum-led staffing policies in effect move from a ratio as an index of staffing policy.

The minimum number of classes that each type of primary school ought to be able to form if a satisfactory curriculum is to be available to pupils. It went further towards defining the curriculum by which its staffing policy should be led. If an adequate middle school curriculum is to be provided... then the staffing must provide some who are capable of specialising in science, craft, foreign language and music. Thus, in theory atleast, staffing allocations, and curriculum-led staffing policies, can be seen as supporting school-based curriculum development. Precisely because, having broadly defined the curriculum provision thought to be appropriate, they highlight the need to identify, develop and exploit staff expertise in order to maintain and renew the local authority, conception of the primary school curriculum. This still obtains as a principle, even where teacher-redeployment strategies adopt criteria other than curricular specialism.

At the present time it is unclear whether decisions about which teachers should be redeployed are based on curricular demands, degree of seniority or local political pragmatism; probably all three are involved to differing degrees. There is no logical reason why redeployment of staff should in itself reduce curriculum development possibilities, assuming that the nettle of curriculum led staffing priorities can be grasped. Put at its most basic this would mean that a teacher with essential specialist curricular expertise would not be redeployed simply because he or she might be the youngest, or the most vulnerable, or the least likely to cause a fuss, or a part-timer.

Equally, it might mean that a teacher with expertise in a shortage area might work in two schools rather than, as is normal, only one. In the recent past the real value of capitation resources allocated to schools has fallen quite substantially, and in some local authorities the reductions were regarded by HMI as endangering educational standards. Although it is a truism to say that such reduction adversely affects the potential of a school staff to improve their curricular provision, the effects will vary greatly according to the nature of the particular curriculum development programme.

There is a great difference in impact of resource expenditure upon a programme requiring for its introduction in a school the purchase of new sets of materials and equipment and one requiring different applications of existing materials and approaches. It has been argued earlier that school-based approaches in primary schools tended to be developments of existing practices, and resource reductions may go some way towards explaining that feature. If so, it suggests that such reductions will influence the nature of school-based development, rather than stifle its growth entirely.

The objective experience of contraction may not necessarily be the most powerful constraint upon development; the ways in which such contraction is perceived subjectively by teachers may be a far more important determinant. Put crudely, curriculum development depends much more upon the ability of a staff to harness its collective energy and enthusiasm than upon marginal reductions in staffing, or even real reductions in resources. It is upon teacher morale that the progress of innovation hangs, and morale itself has been damaged by contraction.

In part, the impact of contraction has been cumulatively routine, with teachers coming in each morning to uncleaned rooms, dusty desks and other evidence of reduced expenditure, which has nothing directly to do with the curriculum, but which effectively eats away at teacher enthusiasm. But the dramatic impact is upon the teachers' perceptions of reduced career and promotion opportunities, which are real enough and may help to discourage innovation amongst career-minded teachers. This is, however, very much a two-edged sword.

Under an expanding system, teachers were sucked up into senior posts by a kind of capillary action and, having obtained their promotion, would demonstrate their merit afterwards. They would be given their special-responsibility allowances and then earn them. In a contracting system such promotion as there is will have to be earned in advance. As a local-authority adviser put it. Previously we promoted Mrs. X in order that she would develop science in a school; now we may be able to reward her if she has developed it. And this has to apply to Scale 1 teachers as well they will have to learn to lead other teachers in an area of the curriculum before they get a responsible post.

Of course the degree of certainty of promotion differed in the two situations. However, the point about shrinking career opportunities is that they too will be perceived differently; some teachers will doubtless be confirmed in their view that there is no point to curriculum development if it is not certain to be rewarded, whilst for others it will at least be a surrogate for, and at best a precursor to, promotion. It would be unrealistic to ignore such a factor in examining the context of primary school curriculum development, and it may be for that reason that a number of recent commentaries have included the notion that it should help to raise morale and channelise the professional energies of teachers who would previously have been promoted more easily. There is, however, a small irony in the situation.

Whatever the general disadvantages of reduced promotion opportunities, from the point of view of school-based development there is one advantage. For development of this kind to be originated, implemented and above all maintained requires, a relatively permanent, or at least stable, staff group. Frequent and widespread teacher mobility is not conducive to school-based approaches to curriculum development.

The impact of the external factors, even in a contracting system, upon a school's ability to develop its curriculum is not overwhelmingly and inevitably adverse. Perhaps the most optimistic element is that primary pupil rolls are expected to stabilise in the mid-1980s and arbitrary loss of specialist skills within a staff group will occur less often. A further issue in the next decade, currently being canvassed in some local authorities and nationally, is the idea that it could become part of the contractual obligation of teachers. If this were to be successfully negotiated it would give considerable impetus to school-based curriculum development, especially if, as is also being considered, it is allied to earmarked resources and a realigned career structure for teachers.

Much of the discussion in the literature about school-based curriculum development is implicitly about secondary schools, so that internal issues raised are problems of relationships between subject departments, of development 'across the curriculum' and of managing large institutions. For primary schools the internal factors pressing upon change are different.

Recognised authority, in respect of the primary school curriculum, is nearly always perceived as residing in the office of headteacher. There is a legal basis for this, but in an age when a broad liberal curriculum in mathematics, English, Science, art, music, craft, physical education and social and moral understanding is expected to be offered to all children, the notion of the headteacher as an authority in all these fields is no longer credible, if it ever was. There is therefore a mismatch between authority and responsibility for the curriculum in primary schools, with teachers who are responsible for developing aspects of the curriculum in which they are the 'expert' having little in the way of formal authority ascribed to them, and the headteachers, who have the authority, lacking the expertise.

Lack of formal status is compounded for postholders by lack of informal recognition from their colleagues, according to HMI, who found few primary schools in which the postholders had a school-wide influence in their subject. It is disappointing to find that the great majority of teachers with posts of special responsibility have little influence at present on the work of other teachers. There are some practical implications of this comment for the primary school post-holder's workload, but the major issue concerns the attitudes of primary school staff towards curricular authority.

Some shift towards a kind of collaborative decision-making about curriculum matters has been proposed by a number of commentators on primary schools, including Coulson and Razzell, although the latter appears, as Harling shows, to understate the obstacles to such an approach being adopted generally. The problem may not be so much about the attitudes of headteachers as those of other teachers. When headteachers delegated responsibility and authority to their deputies, according to Coulson and Cox, the problem was that other teachers did not accept the transfer of authority as legitimate. This brief review of the internal factors has been deliberately uncritical of them in order to report in summary form the prevailing view of the internal culture of primary schools. This internal culture is seen as structuring both teacher relationships in primary schools and teacher perceptions of curricular authority in classrooms and schools so as to provide a major constraint on postholder-led curriculum initiatives development.

## Concept and Theory

The concept of 'primary education' as a distinct area of development was fully established five years before the next report of the Consultative Committee appeared. Certainly the Board did everything it could to keep the concept alive, and after its publication in 1928 *The New Prospect in Education, LEAs* drew up schemes for the full provision of 'post primary' education for children of 11+ along with the lines indicated in the Hasow Report. There is a tendency to look upon it as little more than a historical curiosity; certainly beside the Plowden Report of 1967 it appears a very slim and modest volume, indeed, and although it took two years to produce its gross cost was minute compared with the cost of Plowden.

After a fairly lengthy introduction on general principles, the report provided a historical sketch of the development of the idea of primary education from the beginning of the nineteenth century. It went on to describe the physical and mental development of children between the ages of seven and eleven; and it is interesting to note that, just as Plowden leans quite heavily upon Piagetian thought, the 1931 report used and quoted Padget's *Le Language et la Pensee chez I'Enfant*, published in France in 1923 and translated into English in 1926.

The report then went on to consider the age limits for the upper stage of elementary education, which it felt should be fixed at the age of eleven; that is, the transfer from the primary to the secondary school should take place between the ages of eleven and twelve. It went on to discuss in some detail the arguments against separate 'infant' and 'junior' schools, but finally recommended that, in those areas where it was possible, there should be separate schools for children below the age of seven, and that in all primary schools there ought to be a well-defined line of education between the younger and older children.

The report made it clear that the primary school should not be regarded merely as a 'preparatory department for the subsequent stage' : primary school course were to be planned and conditioned by the specific needs of the child at that particular phase in his development, both physical and mental.

The internal organisation of primary schools was next discussed, including such problems as the size of classes and co-education at the upper stage of primary education. The report adduced evidence from a variety of sources to establish that classes in junior schools should be kept small, preferably about thirty five, and it further made the point that it would be impossible to put into operation many of its suggestions if large classes were retained in primary schools. The size of primary classes was indeed one of the most urgent problems facing education administrators. Concerning co-education the conclusion was that there was no valid objection on general or sociological grounds, provided due regard was paid to the differing needs of girls and boys in games and physical activities generally.

Problems of mental and educational retardation were considered in detail, and the chief causes, detection, diagnosis and treatment dealt with. Retarded children required special attention between the ages of seven and eleven, and it was recommended that special classes should be organised for this purpose and that they should be small.

The traditional curriculum of the public elementary school was next analysed, and the general principles on which the upper stages of primary education should be based were elicited. The report considered the complexity of 'modern industrial civilization' and the bearing this has upon the work of the primary school.

It emphasised the uselessness, as well as the innate danger, of seeking to inculcate what AX Whitehead has termed "inert ideas'; and it deplored and deprecated the fact that, whilst a great deal of teaching was good in the abstract, too little of it directly assisted children to enlarge and vivify their instinctive hold on the condition of life by enriching, illuminating and giving point to their growing experience.

In its further consideration of curriculum detail the report emphasised the desirability of devising new methods on approaching its various branches, and in particular dealt with the project method and 'centres of interest' as set against the traditional practice of treating the curriculum in terms of 'subjects'. It warned, however, against the dangers of such methods when used without due consideration or caution-music and drama were often merely 'dragged in' in order to fulfil what were regarded as the claims of a principle. The staffing of primary schools and the training of teachers were next considered, and it was argued that teachers with general qualification rather than specialists were best suited for work in primary schools.

A brief survey was made of school premises, equipment, school and class libraries, visual and auditory aids to teaching, school visits and playing fields. There followed a discussion of examinations in primary schools and it was recommended that, in classifying pupils leaving the infant school, of consultation between the teachers concerned. Any classification should be merely provisional, and should be subject to frequent revision.

The council felt that as the provision of various types of secondary education was extended in the way proposed in the report on The Education of the Adolescent, the need for selecting by competition those children who would pass on to grammar and selective modern schools would diminish. It was convinced, however, that some sort of qualifying test or examination would always be required for the purpose of classifying pupils, and argued for the use of written papers in English and Arithmetic as a basic test of capacity and attainment of children at the age of eleven, together with carefully devised group intelligence tests. Throughout the report of The Primary School there was a lively sense of the needs of the children themselves. It considered that what any wise and good parent might desire for his own children was precisely what the nation as a whole must desire for all children.

The primary school was on the way to becoming what it should be, namely, the common school of the whole population, so excellent and so generally esteemed that all parents will desire their children to attend it. In 1933 the Hadow Report on the Infant and Nursery school was published and, like its predecessor, it emphasised the need to build new schools for young children more on an open-air plan. It also made it clear that the best place for very young children was in the home; but if this were not possible there was a great deal to be derived from attending nursery schools.

There is a recognition of the child as a total being who has specific and differing needs within the realms of character and intellect, and additionally the need to be initiated into the world of work and life itself in such a way as to be able to cope practically as well as intellectually. Eaglesham may be right when he comments that whilst there is to be preparation for life there is to be no vocational education.

The children are to be trained in habits of careful observation and clear reasoning so that they may gain some intelligent acquaintance with some of the facts and laws of nature and this is hardly 'followership'! The school must further arouse in its pupils a lively interest in man's ideals, achievements, literature and history; it must give them 'some power over language as an instrument of thought and expression'— surely more keeping with education for leadership! All sound education should make pupils and students aware of the limitations of their knowledge and encourage a due sense of humility within the realms of learning-whether they are leaders or followers.

The code consequently adds that, while the school should make pupils conscious of their limitations it should provide thoughtful study as it will enable them to increase that knowledge when their school career is over, and by their own efforts. This, indeed, is still one of the prime purpose of education-to demonstrate to pupils how to acquire knowledge and learning for themselves. The influence in practice of the 'sloud' movement is to be seen in other elements of the introduction which suggests that the school should encourage the natural activities of the child's hand and eye 'by suitable forms of practical work and manual instruction'; and the general, though gradual, movement towards hygiene and health is suggested by the training of pupils in physical exercises, organised games and the simple laws of health.

It is true that the code saw the possibility of discovering 'individual children who show promise of exceptional capacity', and that such children should be qualified to pass at the appropriate age into the secondary schools and there be enabled to derive the maximum benefit from the education offered them. Whilst this was a recognition that some children rather than others were fitted for further education it is worthy of note that there was a clause in parenthesis which had the sort of contention which many educators have put forward in more recent years against the streaming of children in primary schools for the 11+ examination, and the gearing of the whole of the primary curriculum for the benefit of the academic few to develop their special gifts. The introduction was not an exhortation to industry, respect and reverence in quite the same sense as that implied in previous codes and reports.

There was little or no suggestion of 'fellowship' subservience in the terminology of the code. It was implied that there was a great responsibility on the part of the teachers to lay the foundations of conduct; and by their influence, example and personal sense of discipline, which should pervade the school, to inculcate in the children 'habits of industry, self control, and courageous preference in the face of difficulties' training suited, surely, to the middle and upper classed as well as the working classes. Reverence was to be taught, not for the nobility bur for what was noble; and children were to be ready for self-sacrifice and to strive to the uttermost for purity and truth. The respect which was to be fostered was not that for 'their betters', but for others, 'which must be the foundation of unselfishness and the basis of all good manner's. The corporate life of the school was the basis for the development of the instinct for fair play and sense of loyalty to one another-the very 'germ of a wider sense of honour in later life'.

Some of the introduction, for good or ill, reads almost like a public school code. The introduction concludes that school, parents and home should all unite in an effort to enable the children to reach their fullest development as individuals, and also to become useful and upright citizens in their community. Whether Morant was interested in forwarding elementary education or not, there was a certain liberalising and humanising purpose expressed in this introduction. And this was supported by the Blue Book issued by the Board of Education in 1905, which was a handbook of suggestions for teachers involved in the work of public elementary schools.

Uniformity of practice throughout such schools was not desirable, but rather that each teacher should think for himself and work out his own methods according to the school's conditions and requirements. Above all the teacher should know and sympathise with the children he was teaching. The entire process of education was viewed as a partnership for the acquisition of knowledge.

Facts were not to be dealt with in isolation but in relation to the total experience of the child 'each lesson must be a renewal and an increase of that connected store of experience which becomes knowledge. The Blue Book was in line with the introduction to the code in that it insisted that the latter should fully realise his duty to use his innate powers to the best advantage.

Life must be presented as something at once pleasant and serious, and in consequence the work being pursued in the schools was in a real sense a preparation for life. Moreover, the teacher's influence in all this was a very vital one, however short the period of influence might be. The First Education Act of 1918 enforced compulsory attendance at school up to the age of fourteen years, and it also underlined the need for a complete reorganisation of what today is more specifically referred to as primary education, that is, the education of young children below the age of eleven years.

Officialdom had learned a lot from the less orthodox and more progressive movements already mentioned, but despite all this, many of the school buildings were inadequate and unhygienic. And despite the general desire expressed to make the primary school curriculum free from the pressures of selective examinations at the upper end of the school, in practice their classes were geared largely to preparation for secondary school work at the subject level and to some sort of 'scholarship' examination at 11+.

The White Paper on Education Reconstruction which appeared in 1943, suggested in particular that children of 11+ should not be classified on the basis of a competitive test, but rather upon the assessment of their aptitudes based largely upon school records and intelligence tests. It felt that competitive examinations at the age of eleven were wrong in principle, not only because of the strain to which children were subjected but also because the future schooling and careers of children were largely decided at this one point in time. An age when children's minds were nimble and receptive, when their imagination and curiosity were strong and fertile, they were subjected to a cramping and stultifying curriculum in which considerable emphasis was placed upon 'ways and means, or creating the examiners.

The White Paper also stated that there should be separate schools for infants and juniors; and, whilst attendance was not compulsory before the age of five, L.E.A.s must make adequate provision for either nursery schools or, where these were considered expedient, nursery classes in infant schools.

The Butler Education Act of 1944 gave the primary school statutory authority in this country, and it defined primary education in Section 8 as full time education suitable to the requirements of junior pupils'. The Act forced a clear break between primary and post-primary education between primary and secondary education. We are not here secondary education, the developing tripartite system, put ever increasing pressures upon primary schools to act as forcing grounds for grammar schools.

The aims, so clearly expressed in the Hadow Report on The Primary School, were very soon forgotten in the cut-throat competition for grammar school places, and the more enlightened development of primary school methods was somewhat delayed. There was a very real sense in which the firm, separate, statutory establishment of primary and post-primary educational institutions postponed the liberation of primary school methods and the Central Advisory Council for Education was probably not being over-pessimistic when it stated in 1947 that "the gap between a reasonable provision of primary schools and the existing provision was formidable" 'half a century's unremitting efforts will be required before we can hope to have good primary schools for all'. A lot, of course, depends on the connotation of the word 'good' in this context: the verdict of the Plowden Report, twenty years later, was quite simply expressed in these words, 'the primary schools are giving good value for the inadequate amount of money spent on them'. But it made it quite clear that the financial inadequacies has seen severe educational repercussions, as we shall presently see.

In August 1963 the Central Advisory Council for Education was asked by Sir Edward Boyle, then Minister of Education, 'to consider primary education in all its aspects, and the transition to secondary education'. Their report was concluded in October 1966 when it was presented to Anthony Crosland, Secretary of State for Education and Science, and it was published in JANUARY 1967. Volume I presented the report, and Volume II the statistics derived from a national survey of 20,00 schools. The total cost of the report was slightly over 120,000. Part one formed an introduction to the whole investigation and referred in particular to the close association between the home background and academic achievement.

The importance to individual of his family and social background was emphasised, and among question raised in general terms was that of whether 'finding c ut' had really proved to be better than 'being told' and stressed the enormously wide variability in physical and intellectual maturity amongst children of the same age, particularly during adolescence, and the tendency for children to mature physically earlier than formerly.

It went on to discuss the interaction of heredity and environment, the stages of child development as outlined by Jean Piaget, and the measurement of IQ. It found a correlation between children's IQs and parental occupations-the children of professional parents have an average IQ of 115, emphasised that the child was a total personality, and that its emotional, social and intellectual aspects were closely intertwined.

Some of the implications of this section include the fact that individual variations between children of the same age are so great that any class, however apparently homogeneous, must always be treated as a body of children needing individual and different attention. Until a child is ready to take a particular step forward, it is useless to try and teach him to take it. Since any child grows up intellectually, emotionally, and physically at different rates, his teachers need to know and take account of his 'developmental age' in all three respects.

The child's physique, personality and capacity to learn will develop as a result of continuous interaction between his genetic inheritance and his environment. Whilst the genetic factors are not as yet under our control, the environmental factors largely are Part of the report, therefore, suggested the need of a very personal approach to the pupil in the primary school, with a special study of each child's 'readiness' for any particular form of learning or operation; and, at the same time, the provision of a sound, healthy, amenable sort of environment in order to present the best learning situations and conditions.

A minimum programme was suggested for the participation of parents in the education of children a welcome to the school, meetings with teachers, open days, information for parents via brochures about the organisation of the school, and reports for parents. There should be a concerted effort that the primary school should be used as fully as possible, out of ordinary school hours, as a sort of community centre. Community schools should be developed in all areas, but particularly in educational priority areas, where there was need for constant communication between parents and teachers if the schools' aims were to be completely understood. The report emphasised the need for colleges of education to have stronger and more efficient links with such schools and to develop courses to meet the needs of immigrant children in particular.

There was a very strong suggestion that the training of teachers generally should take more account of those social factors which affect school performance, and also of the structure and functions of the school services. It recommended the initiation of experimental schemes in the joint training of teachers and social workers, and already there are colleges of education which are devising courses for social and youth wing work as well as general welfare work. The report made it very clear indeed that the primary school has a social role in the community, as well as an immediate and individual addictive role.

There should be three years in the infant school and children should not be transferred until the age of 8 years; this would permit both children and teachers to work steadily and without anxiety. The age for admission to a secondary school was suggested as 12+, since this would give a four-year course in the junior or middle school, with a median age range from 8 years 6 months to 12 years 6 months. It should be noted here that the definition of primary education provided by Section 8 of the 1944 Education Act was amended by Section 3 of the 1948 Education Act.

Primary education was there defined as 'full time education for children below 10 years 6 months and children above that age but below 12 years who it is expedient to educate with them'; whilst the 1964 Education Act allowed proposals to be submitted to the Secretary of State for the establishment of new schools with age limits below 10 years 6 months and above 12 years. The report emphasised the need for the fullest possible documentation of each pupil before transfer to the secondary school, and it detailed the type of contents that a folder on each child should contain.

The ill effects upon primary education of selection for secondary schools were lessening, and the council recommended that authorities still employing selection procedures should no longer rely on an externally imposed battery of attainment and intelligence tests. Part five dealt with the children in the schools, with curriculum and internal organisation. Among a large variety of aims and purposes mooted, and dangers to be avoided, the following aims of primary education were accepted in somewhat general terms, with the provision that generalities have limited value and can quickly become little more than platitudes; it was agreed that a pragmatic approach to the purposes of education was more likely to be fruitful.

1. Adaptability-to fit children for the society into which they will grow up, and to train them to be capable of adjusting to their changing environment.
2. The all-round development of the individual child.
3. The acquisition of the basic skills necessary in contemporary society.
4. The religious and moral development of the child.
5. Physical health, intellectual development, emotional and moral health, aesthetic awareness, a valid perspective, practical and social skills, and personal fulfilment.
6. Values and attitudes must be mediated to the children. The school is not merely a teaching shop but a living community in which pupils learn primarily to live as children and not as future adults.

The work of Piaget was emphasised as a sound developmental approach to children's learning, and the importance of children's 'cultural' play was made clear. The report went on to consider certain particular aspects of the curriculum, and throughout there was an enlightened approach to both content and method, and a balanced attitude towards heuristic principles.

The council has no doubt that 'children's questions about sex ought to be answered plainly and truthfully whenever they are asked'. It gave validity to the modern, both relaxed and friendly, approach within the primary schools as a much better preparation for life in contemporary society than the old authoritative one.

Whilst accepting "discipline", the council were clear that this connoted neither heavy punishment nor soft and flabby relationships; discipline, it felt, was impaired by such elements as disorder, untidiness and slackness. It could flourish only in an ethos of order and purposefulness, in which boredom had been eliminated. There must be, in all this, a healthy combination of individual, group, and class work and learning. Due consideration was also given to the education of both handicapped and gifted children. The role of the teacher was discussed, and it was argued that teachers must enlarge their endeavours and enlist to a greater extent parents' interest in their children's education. It was extremely important to diagnose the child's needs and potentialities and there would be increasing demands on the knowledge of the teacher, whether literary, scientific or mathematical.

Teachers said the Report 'cannot escape the knowledge that children will catch values and attitudes far more from what teachers do than what they say unless they are courteous, they cannot expect courtesy from children: when teachers are eager to learn and turn readily to observation and to books, their pupils are likely to do the same'. The report, in effect, asked not only for more teachers but also for better quality ones. It went on to discuss teachers' aides, or trained ancillaries, who might give substantial help to teachers inside and outside the classroom, and who would have equal status with nursery assistants and have comparable training.

The council considered that colleges of education, in general, were too remote from the problems of the school; it recommended a full enquiry into the system of training of teachers, and suggested that there should be more joint appointments to college and school staffs. Further, a network of residential teachers' courses should be developed.

1. The Department of Education and Science should consider taking steps which would require all independent schools to state on their prospectuses whether the schools were recognised or registered and what this implies. The

Department of Education and Science should reconsider the terms "recognised" and "registered" and try to devise more informative ones.

2. The Secretary of State's powers to serve Notices of Complaint on independent schools should be based on more stringent criteria. The construction of "objectionable" should be widened to include any conditions, physical or educational, in which children's welfare was not thought to be adequately safeguarded.
3. All head teachers of independent schools should be qualified teachers. After a data to be specified, only qualified teachers should be appointed as heads in new schools, or when there is a change of head teacher.
4. In-service courses should wherever possible allow some places for teachers from independent schools. The independent schools themselves should take through their professional organisations to "increase the facilities for inservice training for teachers in independent schools."

The primary schools were giving good value for the inadequate amount of money spent on them, and went on to state that teachers were too few in number and too unevenly distributed to do the job adequately.

The first priority was the establishment of education priority areas; the recruitment of teachers' aides would be an essential and immediate source of help to the schools everywhere, the essential improvement of bad primary school buildings must be undertaken as soon as possible wherever they existed; and nursery education must be extended through increased accommodation and staff. Planning should begin on changes in the national dates of entry, and also on the ages of transfer between the different stages of primary education.

The Plowden Report represents one of the most thorough investigations into any area of education ever produced. Perhaps not least of its achievements was its virtual unanimity, which is reflection on the goodwill and singleness of purpose of its twenty-five members over a period of three years of intensive work and study. Out of every 100 women who enter our colleges of education, only 47 will be in the schools after three years of teaching; and after six years of service only 30 will remain. Perhaps the most sanguine way to look at the figures is simply to argue that 70 women out of every 100 so trained have, at least, qualified in a profession to which they may later return, but in any case they have found one way of obtaining higher education, with often minimum of entry qualifications; and as mothers, in the majority of cases, they will have a better understanding of educational problems.

8

# Planning the Lessons

In modern system of teaching, effective lesson-planning depends upon various factors. The knowledge of the physiological development and the intellectual maturity of the pupils is essential before making a lesson-plan for the class. Similarly, a knowledge of the needs, interests and abilities of the group helps him to make the lesson-plan suitable for the group. Moreover, the knowledge of the psychology of learning, principles of teaching, previous knowledge of the pupils of the class, effective mastery of the subject matter, etc., are essential prerequisites of lesson-planning. The teacher who prepares the lesson-plan should realise the importance of the various steps of a lesson-plan. It demands sufficient experience of the teacher to plan classroom activities to develop understanding, interest, aptitude and skill of the pupils in addition to the scientific knowledge in all its aspects—scientific terms, facts and principles, ideas and concepts, etc. In a lesson-plan, there should be scope for creative activities by the pupils and should provide opportunity for critical thinking.

Questioning is an art; it requires experience on the part of the teacher to frame proper and adequate questions which will lead the pupils to think. A lesson-plan, however carefully prepared, will not serve as an effective teaching tool unless the teacher knows how to conduct the class, even departing a little from the set plan (but still based on it), in order sustain the interest of the pupils throughout the period. The science teacher should be able to plan for the duration of the time available because a science lesson invariably includes practical demonstration or practical work by the pupils which requires much time. Hence, required a sense of timing and management of time.

## Classroom Lesson Planning

Today, classroom lesson planning is essential for successful teaching. A lesson-plan gives direction to the teacher and checks haphazard teaching. It is a classroom guide for the teacher as a visualisation of the classroom experiences he desires to occur. The Dictionary of Education (Good) defines a lesson-plan as a teaching outline of the important points of a lesson arranged in the order in which they are to be presented; it may include objectives, points to be made, questions to be asked, references to materials, assignments, etc. Lesson-planning calls for intelligence, skill, ability and experience of the teacher. A lesson-plan reveals the approach of teaching and also mastery of the teacher over his subject. It, however, does not mean that the teacher has to follow it slavishly; he should use the lesson-plan as basis for developing the lesson through flexible instructional techniques.

There is danger in depending solely on the lesson-plan. Some unforeseeable event may occur in the classroom which may upset the previous planning of the teacher. In such cases, the teacher has to be resourceful enough to manage the situation; an unskilful teacher is sure to be in trouble in such a situation. A lesson-plan should, therefore, be conceived as a guide to effective teaching. It is advised that the lesson-plans should be brief and include the broad points only so that the teacher can treat the details with freedom and according to the circumstances demand.

The teacher should come well prepared so that any departure from the pre-planned arrangement does not cause any difficulty. Lesson-planning is a device essential to economise time and to make the teaching systematic. It enables the teacher to aim at objectives appropriate to the lesson and suitable for the class and thus guides him in the attainment of those objectives.

A lesson-plan enables the teacher to cover a wider field in a limited time. Since the questions or assignments are previously planned, they are thought-provoking and relevant. The important questions relating to the scientific topic and the application of science to our everyday life find relevance as planned rather than extraneous questions. The teacher feels confident with the lesson-plan by his side, because he is aware of the difficulties that may arise in the classroom and is prepared to deal with them.

A lesson-plan reminds the teacher of the specific goals to be attained through classroom teaching and gives him a clear picture of what is to be pursued and what is to be avoided. Planning a science lesson for the class requires selection and organisation of the subject matter of the lesson for presentation in the class. He also has to select the teaching aids and materials such as apparatus, equipment, charts, models, diagrams or films, etc., before giving the lesson.

Moreover, it requires him to plan activities for demonstration or individual and group practical work. It also provides for planning activities to develop the other virtues peculiar to the teaching of science. Blackboard summaries of the essential points, class tasks and home assignments are other important features of a lesson-plan.

A lesson-plan enables the teacher to connect the present lesson with the previous lesson and the lesson that will come next and thus to maintain a continuity in teaching. Teachers should indicate in the lesson-plan the reference materials such as books and journals to be read in connection with the lesson and also arrange for field trips or visits if necessary which may increase the effectiveness of learning. Adjusting the lesson to the allotted time is a problem to be solved by the individual teachers. A lesson should be planned immediately after the preceding lesson, because the events of the previous lesson are fresh in the mind of the teacher.

The extent of content covered and approach followed is distinct in his mind. He also remembers the mistakes he committed so that in planning the next lesson, he can use his experience. He can also foresee the type of activity that will create interest and enthusiasm in his pupils.

## Content of Lesson-plan

### *Statement of Objectives*

The objectives of teaching the particular topic of science should, first of all, be stated in clear, unambiguous words. The whole structure and approach of the lesson-plan will depend upon the objectives desired to be attained. Aims are too general and are directly related to the general aims of education. These should be remembered by the teacher rather than stating them. Objective may be divided into two categories—general objectives and specific objectives. General objectives are of general nature and cannot immediately be developed by the experiences of the particular lesson.

They may be considered as reference objectives for the lesson. Specific objectives are those which can be developed and measured in the classroom. These may include the knowledge, skill, habits or attitudes which can be directly attained in the classroom. For example, in a lesson on thermometers, one of the general objectives may be "to develop in the pupils scientific attitude and spirit of enquiry" of the world around them; and one of the specific aims may be "to enable them to construct and graduate a thermometer." Similarly, in a lesson on dynamo, one of the general objectives may be "to develop in the pupils an understanding how energy can be transformed" whereas one of the specific objectives may be "to help them to know how a dynamo works." The specific objectives lead to the attainment of the general objectives.

### *Apparatus and Aids*

The apparatus and other teaching aids needed in connection with the lesson should be mentioned. A study of the content and the method will enable the teacher to decide the materials needed. He will, however, have to consider the available resources and facilities. The selection of apparatus for a demonstration or laboratory work and related reading materials or references, charts, diagrams, models, films or filmtrips, chemicals will depend upon their availability.

### *Preparation (Introduction or Motivation)*

Before giving the lesson the teacher needs to know the pupils' previous knowledge. This is to know the background of the class and thus to prepare ground for developing the present lesson. Through proper questioning, the teacher should test the previous knowledge of the pupils and then introduce the lesson linking it with their previous knowledge. There may be various ways of introducing the lesson. It may be introduced through an interesting demonstration or telling of an interesting story of the discovery of the particular fact or principles or the biography of the scientist connected with it. Appropriate questions may be asked through which the pupils may be led to realise the necessity of learning the new topic the teacher is about to present. Arriving at this stage, the teacher should announce the day's lesson. After introducing the lesson comes the stage to present the lesson.

### *Presentation*

This is the stage of actual giving of the lesson, the time for the teacher to show his ability in selecting, organising and presenting the content matter, engage himself in appropriate activities, such as demonstrating, talking, questioning, supervising, giving blackboard summary, or simultaneously evaluating and also providing activities to the pupils such as observing, taking notes of the salient features, helping in demonstrations, doing individual or group practical work. The pupils should be kept busy with one or the other type of activity. During teaching, the teacher should encourage the pupils to observe carefully, compare and contrast with similar events and lead them to generalise the principle taught in the class. Comparison of the present features with similar other facts and principles is essential in any type of scientific learning. The arrival at a generalisation by the pupils may not always be completely correct, yet it helps to attain the broader objectives of teaching the particular topic of science.

### *Application*

In this stage, the application of the new learning to life situations is discussed. The pupils should be able to see the application of principles or the phenomena in their everyday life and also be familiar with their applications. They should be able to apply their knowledge of facts and principles to new situations. They may be asked to solve problems which involve the application of the facts or principles taught in the class. Tests may be applied or questions may be asked to test their knowledge or some creative exercises may be given where the pupils will have to apply their knowledge. This stage may also be used for the purpose of revision or recapitulation.

If any activity is well planned in advance comes out to be successful, so is the classroom teaching. If a teacher plans his lesson beforehand, he knows what he is going to teach, how he is going to teach, what teaching aids he is going to use, what questions he is going to ask, what questions students might ask, what objectives he is going to achieve and how he is going to evaluate that he has achieved his identified objectives. The answers of such questions make a lesson plan. For a beginning teacher it (lesson plan) may be a detailed script and for an experienced teacher it may be a small package of some teaching points in the written form or just in his mind. For a science teacher a lesson plan comes out to be very effective and economical.

It is advisable for a science teacher to select and organise in a logical sequence, the suitable material for classroom instruction. Where should he start today's lesson with respect to where he ended yesterday? If yesterday's lesson ended with the completion of a demonstration showing that certain solutions under given conditions conduct electricity, how should he develop his lesson today? Should the demonstration be repeated at the very beginning of the lesson? Should the students be asked to tell what they observed yesterday? Should the same apparatus for conductivity of solutions be placed on the demonstration table. Should the teacher state the conclusions drawn, yesterday? Should the basic generalisations be asked from the students?

When teaching a certain lesson, what questions should the teacher ask? Should the teacher answer all the questions raised by the students, or the students leave the classroom with some questions unanswered? When the teacher is lecturing or demonstrating, should the student note down all the information, applications or uses?

There are several possible answers to the above questions. The teacher should give serious thought to such questions when developing his lesson plan, in order to arrive at an efficient and economical approach to teaching. The format of the lesson plan and the actual written material in it are not of great importance. The significant factor is that the lesson plans provide the science teacher with the opportunity of checking which of the suggested teaching methods are potentially most effective for achieving the identified objectives of the lessons.

## Challenging Learning

In addition to the four criteria suggested above for lesson planning, the science teacher should be conscious of some other important points when he is just about ready to teach. Why should these students desire to learn the solution to the problem or understand the scientific principle? Psychologists often refer to *motivation.* How does the teacher motivate the students to learn each lesson? The teacher can set the stage for learning by presenting an inspiring demonstration, a series of experiments, and questions or problems in which the students can become involved or in which they can identify themselves with specific situations.

Students interests, needs or experiences can serve as the springboard for introducing the lesson. Some teachers may begin with their own personalised experiences, the showing of a film or slides, or having a student or group of students bring in and discuss a science project. There are some typical ways in which the science teacher attempts to stimulate learning with the hope that the students will be motivated to learn. Here lies the need for flexibility in the lesson planning.

Some students may actually assist in the planning of instruction if given the proper opportunity. Such students should be more strongly motivated to learn the basic ideas of the lesson under the guidance of the science teacher, so that they could assist their teacher in the classroom when teaching science to the entire class.

## Stages Involved in Lesson Layout

There are many types of lesson plans. The Hebertian steps provide us with the best, basic lesson plan. Hebertian steps of a lesson plan are as follows:

1. Preparation.
2. Aim.
3. Introduction.
4. Method (Presentation).
5. Application.
6. Recapitulation (Revision and Conclusion).

### *Preparation*

However, knowledgeable you may be, and however certain you may be of what you wish to teach, you must plan how much and by what method you propose to teach for a successful lesson.

Time must be your first consideration. Very often the bell rings when the lesson is just half completed. Duration of period should be divided into three parts — time for teacher's activities, time for students activities, and time for recapitulation.

Secondly, you must plan your lessons in terms of the specific objectives for the lesson identified by you, i.e., what you wish to achieve in a 30-40 minutes period at your disposal.

Thirdly, you must plan your lesson according to the cognitive level of your students, taking into consideration their age (maturity level) and aptitude.

Fourthly, you must well prepare yourself for the lesson asking yourself a few questions like these. Do you have the requisite knowledge and background for the lesson? Can you perform adequately the skills you propose to teach? Have you considered the way you will start and proceed for, your lesson?

Fifthly, you must prepare all the materials and teaching aids necessary for the lesson, asking yourself a question. Have you arranged these materials (specimens, charts, illustrations, maps, diagrams, models, etc.) in a sensible manner so as to achieve sequence and unity in your lesson.

Sixthly, you must ensure that your students are prepared before the lesson starts with everything necessary for the lesson.

It has been seen infinitely that more lessons are ruined by neglect of these simple preparations. It would be advisable that a sound teacher should consider himself as a stage manager who must have all that is necessary in the way of effects, scenery and properties available at the precise moment that they are needed so that a smooth production is achieved.

There is, moreover, a vital psychological aspect to this question of careful preparation. Students react quickly to efficiency. If they are obliged to behold daily a fumbling an ill-prepared teacher they will speedily lose respect for such a person. On the other hand the daily spectacle of business-like preparation on the part of their teacher will earn their admiration, and incidentally, their polite attention.

***Remember:***

Prepare your lesson.

Prepare yourself.

Prepare your materials and teaching aids.

Prepare the material and aids for your students.

### *Aim*

Following a successful preparation the teacher must state clearly the aim of the new lesson. (The aim may further be subdivided into some general and specific objectives which are going to be achieved during the lesson.) The students must know the purpose and objectives of the lesson. If they know clearly what they are supposed to achieve, then their thought processes will follow more easily the teacher's exposition, explanation and steps of the lesson.

***Remember:***

A class of students without an aim is like a leaderless army.

The students must know where they are going.

The aim of the lesson must be stated clearly and emphatically.

### *Introduction*

If the lesson has been sensibly prepared, the teacher will have taken pains to plan a suitable introduction. The introduction should not only set the atmosphere or mood for the lesson, but it should gather together the students and focus their concentration on to the subject in hand. To prevent initial in attention, attention should be paid to two aspects of the introduction stage of your lesson — social introduction and subject introduction.

### *Social Introduction*

You should greet the students courteously and discuss with them the exciting things they have been doing. Be human and sociable with them for a minute or two and gradually calm them down by friendly conversation.

### *Subject Introduction*

After this brief social introduction, you should introduce them to the nature of the present lesson. This is perhaps best achieved by reminding them (by questioning) of the ground covered in the previous lesson on the subject, or activity in hand. Educationally it is vital to proceed from the firm base of the 'known' before attempting to explore the territory of the 'unknown.' This can be achieved by putting the new knowledge which is linked up with their previous knowledge gained in the previous lesson or from everyday experiences.

In order to develop curiosity for learning, the teacher should make the students realise that their knowledge is incomplete and a lot of new knowledge is still left to be acquired by them. Thus, you may involve the following steps:

1. testing their previous knowledge;
2. asking short reflective questions enabling students to realise their ignorance of numerous scientific facts and create interest therefore in learning the new subject matter;
3. using audio-visual aids; and
4. performing interesting demonstration relevant to the topic of the lesson.

***Remember:***

Introduction is the bridge between the previous activity and the present activity.

Introduction creates the new mood for the new lesson. A introduction is necessary to prepare the minds of the students for me new topic of activity. Introduction is the process of rallying the minds and bodies of children to face a new experience.

***Method (Presentation)***

You have learned several methods of teaching science. You should practise all the methods in your practice teaching taking into consideration the:

1. objectives of lesson,
2. cognitive level of students, and
3. classroom conditions.

When teaching science it should be kept in mind that students' activities are predominant than the teacher's activities. Remember it is the student's lesson, and not your lesson. Restrict yourself to a brief introduction, statement of aim and a short, thought-provoking exposition and explanation and then employ to the full all the senses of the students (as for as possible) — the hands (learning by doing), the eyes (learning by seeing), the ears (learning by hearing), the mouth (learning by saying), the nose (learning by smelling), and the tongue (learning by taste).

Only at the end of the lesson you should use your voice emphatically to summarise the lesson. If your students find out things themselves and reach solutions by constructive reasoning there will be infinitely more satisfaction for them than if they are constantly 'told' by you.

Thus, after the students are prepared to receive the lesson, and the teacher has clearly stated the aim of the lesson, he should start the lesson, and the students should get the new ideas and knowledge. Here the teacher should draw out the maximum from the students by asking suitable questions.

The lesson should proceed with the actual participation of both the students and the teacher. It can be made more interesting and concrete by the use of the suitable audio-visual aids and experimental demonstrations.

The new ideas and knowledge to be given to the students should be related as far as possible with their previous knowledge. This is a very important point to be kept in mind when introducing new scientific principal and generalising some new scientific facts.

Finally the students are in a position to generalise what they have learned so far. Answers to some relevant questions now asked by the teacher, he will be able to draw out various scientific principles/ conclusions and formulae.

Thus, the students receive the new knowledge and the teacher acts as a guide for them.

***Remember:***

Your methods of teaching should a make the students think a make the students. Do a make the students. See a make the students. Say a make the students hear.

***Application:***

In order to develop interest and satisfaction among students for the subject they should be made familiar with the use and application of various generalisations in daily life and conclusions drawn out. The new knowledge thus gained by the students becomes more meaningful and permanent for ever. In former times, application merely consisted of the students proving by stereotyped exercises that they had learned what teacher had taught them.

The very words 'test' or 'exercise' explain this former conception. Without denying the value of this practice, we find that application has, to the progressive teacher, a much wider significance.

As we have decided, it is vital that the teacher employs an interesting, imaginative and lively method of presenting his subject matter, it is equally vital that such a presentation be matched by an interesting, imaginative and lively application.

***Remember:***

Knowledge and skills must be applied and practised if they are to be lasting benefits to your students.

Minimum of presentation, maximum of application.

Match an interesting presentation by an interesting application.

***Recapitulation (Revision and Conclusion)***

"Teach little, revise more" is an excellent slogan for a teacher. It is not correct to imagine that more than sixty per cent of anything you teach for the first time will be observed by the students. For this reason, you must reinforce what you have taught by causing the students 'to see, say and do ' the things you have presented to them orally.

Revision is essential to any piece of teaching and without it the ablest presentation will only be partially successful. You must revise as you teach and you must revise as you have taught.

It is vital that no lesson should be concluded without an emphatic summary of what has been taught. At least five minutes must be allocated by the teacher in the lesson plan for this purpose of hammering home what has been taught, by a systematic review or recapitulation of what has been achieved.

Here the knowledge imparted to the students is tested by asking some relevant questions. This is a way of revising the lesson by question-answer technique. This also enables the teacher to find out whether the students have followed what he has taught them, and whether or not the method of teaching is effective.

This summary should be first drawn from the students by questioning and then repeated with vehemence by the teacher as he points to the main items summarised on the blackboard. Revision is thus oral, aural and visual. The teacher speaks and illustrates and the pupils hear and see.

***Remember:***

Revision, Revision, Revision.

Now Summarise and Revise the Stages of Your Lesson Plan.

You have *prepared* your lesson thoroughly.

You have stated your *aim* clearly.

You have *introduced* it humanly.

You have employed an interesting *method*.

You have devised an imaginative *application*.

You have *revised* your lesson decisively.

## Unit Layout

Planning a Unit is similar to lesson planning. A Unit may have several lessons. Therefore, it may not be completed in one class period. It may take several class periods to complete a Unit. If Unit Plans are given in Teacher's Guides, they might be very useful for the science teachers specially the new teachers.

A new teacher feels difficulty in treating a topic in his class not because he does not know the subject matter, but because he does not know exactly how much subject matter in a certain topic suits the varying age-levels of the students. Again he faces the difficulty in selecting the teaching material. He also finds it difficult to decide what activities should be undertaken to cast a lasting impression on the minds of the students. Very often, the teachers specially the new ones do not have enough practice of how to evaluate how far the students have followed what they have been taught.

But for the many experienced science teachers it may be quite easy. Thus, a beginning teacher may take advantage of their experiences, but they are not always available to him. But if NCERT and SCERTs invite such experienced science teachers to develop Unit Plans on some selected science topics, it would be of great help not only to freshers, but to others too. Such a 'Unit Plan may also be termed as Teaching Unit.'

***Steps of a Unit Plan***: A Unit Plan may be written in the following steps:

1. Subject: (Physics, Chemistry, Biology or Science)
2. Topic: (Heading of the Unit)

3. Class: (To whom the Unit is to be taught)
4. Time: (No. of class periods)
5. Aids: (To be used during the entire unit)
6. Content: (Syllabus to be covered)
7. Sub-units: (No. and Name)
8. Objectives: (To be achieved after completing the Unit)

For ***Each Sub-Unit:***

9. Sub-unit No. and Name (Heading)
10. Time: (No. of class periods)
11. Teacher's Activities
12. Student's Activities
13. Joint Activities (Teacher's and student's)
14. Assignment
15. Follow-up Activities
16. Evaluation
17. References

Under these steps A Sample Unit Plan "Lifting of Heavy Bodies by Smaller Forces" for IX Physics is developed and is given here for guidance. Similar Unit Plans may also be developed in other science subjects — Chemistry or Biology and for any class level.

***A Sample Unit Plan:***

*Subject:* Physics

*Topic:* Lifting of Heavy Bodies by Smaller Forces

*Class:*IX

*Time:* 11 periods, each of 30-40 minutes duration

*Aids:* Tool chest. Planks of wood, 16 mm film projector/ VCR/VCP.

***Introduction*:** Here a teacher is not to teach all that is given in the chapter on 'simple machines' found in any secondary physics textbook. What is important in the class level we are going to teach, i.e.; the intellectual level of me learners.

1. Intricate mathematical calculations should be avoided.
2. Subject-matter not in keeping with the cognitive level of learners should be eliminated.
3. A teacher, however, may add something new that he deems necessary, even if it is not given in the textbook.

This Unit may be divided into the following sub-units:

1. Classification of different devices
2. Levers .
3. Wheel and axle, and pulley
4. Inclined plane, and jack.

***Objectives:***

1. To find out from children what devices they have used or seen being used for lifting heavy bodies.
2. To classify these devices into different categories.
3. To help the children understand how these devices are used in everyday living.
4. To help the children construct such devices or their models.
5. To enable the children to distinguish between the utility of the various devices in various situations.

***Sub-Unit 1: Classification of Different Devices:***

***Time:*** 2 Class periods (60-80 minutes) *The Teacher's Activity:*

To pose the following problems:

1. How can a child defeat a team of wrestlers in a tug of war?
2. How can a man have his truck lifted in order to remove a flat tyre?
3. How could a new born baby lift his father?
4. How can a man lift his scooter on a platform, which is about one metre above the road?

***The Student's Activity:*** With the help of past experiences the students will:

1. try to solve these problems, and
2. suggest why certain devices are used and not others. (These devices will be listed on the blackboard).

***Part Played by the Teacher:*** The teacher will give finishing touches to the list of the devices suggested by the students. He may ask the students to bring, either the actual devices or the models to the next class meeting. The teacher's own enthusiasm and initiative in this direction will be of supreme importance.

***Joint-Activity (Teacher's and Student's):*** The following day, the devices brought by the students the actual ones or in the form of models should, after demonstration and explanation, be classified into:

1. Lever;
2. Wheel and axle;
3. Pulley;
4. Inclined plane;
5. Jack.

***Sub-Unit 2: Levers***

***Time:*** 2 Periods (60-80 minutes) *Main Activity:* Projection of 16 mm film on 'Levers' showing its video:

1. What is a lever and how is it used?
2. Experiments with levers.
3. How often levers are used in everyday life?

***Teacher's Activity***

1. The teacher should see the film beforehand. If the commentary is in English and the language is too difficult for the students, the teacher should prepare his own commentary (in Hindi or any other regional language) and give his commentary stopping the sound of the projector.

2. Pre-projection activities should include the introduction of the topic to be viewed by the students, so that they should know what they are going to learn from the film. Follow-up activities should include clarifying concepts not followed by the students.

*Assignment:* Classroom activities may be coordinated with science club activities. Students may engage themselves in making models and drawings of levers. A time-limit for the completion of the work should be set up by the teacher,

***Sub-Unit 3: Wheel and Axle and Pulley***

*Time:* 2 periods (60-80 minutes)

*Main Activity:* Projection of 16 mm film on "Wheel, and Axle, and Pulley" or showing its video:

1. The construction of the wheel and axle device.
2. How this device increases force and speed?
3. In what situations pulleys are used?
4. Types of pulleys.
5. How a small child can defeat a strong man in a tug of war by using pulleys?
6. How all these devices help us in everyday life?

*Teacher's Activity:* Same as in Sub-unit 2.

*Assignment:* Similar to Sub-unit 2, to suit what is taught in Sub-unit 3.

***Sub-Unit 4: Inclined Plane and Jack***

*Time:* 2 periods (60-80 minutes)

*Main Activity:* Projection of 16 mm film "Inclined Plane, Wedge and Screw" or showing its video:

1. What is an inclined plane and how is it used to increase force.
2. What is a wedge, and how is it used in everyday life?
3. What is a screw, and why is it considered a machine?

***Teacher's Activity:*** Same as in Sub-unit 2.

***Assignment:*** Similar to Sub-unit 2 to suit what is taught in Sub-unit 4.

***Follow-up Activities***

***Time:*** 2 periods (60-80 minutes)

1. A display in the laboratory of the charts and models made by students.
2. Those who have made the models and charts could be asked to explain their work to other students.

***Note:*** This activity will need the supervision of the teacher.

***Evaluation:***

***Time:*** 1 period (30-40 minutes)

The construction of an objective type test on the matter covered, and to see if the objectives of the Unit have been achieved.

***References:***

1.16 mm films

(a) Levers — English, 11 minutes

(b) Wheel and Axle, and Pulley — English, 11 minutes

(c) Inclined Plane, Wedge and Screw — English, 11 minutes.

**Note:** These films or videos can be borrowed from The Central Institute of Educational Technology (CIET), NCERT, New Delhi. 2. UNESCO Source Book of Science Teaching.

## Planning the Lessons

The important features of what a lesson plan should contain are:

1. Why should we teach this lesson (Objectives)?
2. What should we teach in this lesson (Content)?
3. How should we teach this lesson (Methods)?
4. How should we know that students have learned what we taught them (Evaluation)?

### *Objectives*

In listing objectives for a particular lesson, careful thought should be given to such categories as understanding factual information and its application as well as developing: (a) problem solving skills, (b) scientific attitude, (c) critical thinking, (d) interest in science and (e) values. It is unlikely that objectives covering all categories be attempted in one lesson. Some of these objectives like developing scientific attitudes or problem solving skills may be attended over a long period of time after several lessons.

Such objectives should be written as general objectives. When developing lesson plans, the teacher should write only those objectives (as specific objectives) for which he is quite sure, that he would be able to achieve at the end of that lesson. Sometimes some student-teachers write quite a number of objectives in their lesson plans, perhaps to impress their supervisors, while only few of them they probably achieve.

### *Content*

The teacher should take up only that much content, which he feels he will be able to cover in the limited time. In introducing a lesson, especially when a new unit is being taught for the first time, the teacher should attempt to stimulate, motivate and generate student's interest in the topic. Student's past and present experiences can serve an effective approach to new units.

Other ways of launching content with the ultimate aim of motivating students to learn a particular topic may be: (a) giving special homework assignments that require research, (b) survey conferences, (c) investigations, (d) special readings, (e) experiments, and (f) observations. The teacher may also bring in or ask students to bring in new articles (from magazines, journals or newspapers) of contemporary importance and relate such reports with the content of the lesson as well as with the needs and interests of the students and the community.

Students may also be encouraged to bring in various supplementary material pertaining to the topic of the lesson.

In short the teacher should utilise all these means for the selection of content and should not limit himself only to the textbook.

### *Methods*

There are several methods of teaching science. Now when the teacher has identified the objectives of the lesson and selected the content to be covered, this is the right time to select an appropriate method of teaching the lesson. He should choose the method/methods which best fits/fit in the lesson. The teaching method should be compatible to the cognitive level of students, identified objectives and existing classroom conditions (number of students, size of the classroom, duration of period, availability of apparatus for classroom demonstrations and experiments, and other audio-visual aids, etc.). There are several audio-visual aids for teaching science.

The teacher can select some appropriate audio-visual aids from those available and improvise or make some on his own which best fit in the lesson. It should be noted that only appropriate-audio-visual aids should be used in the lesson, and those too if they are actually needed. Sometimes mostly student-teachers use so many audio-visual aids in one lesson, which make the lesson more confusing instead of interesting and clarifying the difficult ideas. This should be avoided. The appropriate audio-visual aids may be used to introduce the lesson as well as during the entire lesson.

### *Evaluation*

This part of the lesson plan helps the teacher to determine to what degree learning occurs during the lesson. Several key questions asked by the teacher (during the lesson or at the end of the lesson) and the kinds of responses obtained from the students will enable the teacher to evaluate the degree of learning. Students' questions will also give the teacher some insight into the nature of difficulties that some students might have. A summary which elicited from the students at strategic places during the lesson as well as at the end of the lesson will also provide the science teacher with an evaluation of the teaching-learning process.

## Skill for Asking

The art of questioning is also critical when teaching a lesson of science. We should not underestimate the importance of questioning skills increasing or diminishing student participation. We ask questions for a variety of reasons.

1. To obtain feedback about the level of understanding in the class.
2. To know what they already know.
3. To promote thinking.
4. To draw attention to something.
5. To provide the student with an opportunity to verbalise his ideas in a coherent way.
6. To give opportunities to praise and encourage a student, and
7. To act as a measure of class control, etc.

To meet this variety of purposes requires many forms of questions, some of greater subtlety than those which have one simple, factual, and correct answer. To promote thinking requires more open ended questions, i.e., those having more than one acceptable answer.

In addition to questions and students' oral responses, evaluation in the course of a period may also be made by observing the students working at their seats or those who may be asked to come to the blackboard. When students are in the science laboratory careful observation of their work habits will also reveal the nature of their learning. A science teacher can observe a student's error at the board, at the desk or in the laboratory and correct the mistake promptly to prevent a block to learning the major and minor ideas of the lesson.

# Bibliography


Abdi, W. H.: *Interaction between Indian and Central Asian Science and Technology in Medieval Times*, Concept, New Delhi, 1990.

Abraham, M. F. and Abraham P. S.: *Science: Development and Change,* Hall Press, Wyndham, 1988.

Addis, Ababa: *Education and Training in Science and Technology in Asia,* UN Economic and Social Council, New York, 1989.

Adiseshiah, Malcolm S.: *Science and Technology for People,* Tamil Nader State Council for Science and Technology, Madras, 1985.

Anderson, R. D.: *Developing Children's Thinking Through Science,* Prentice Hall, New Delhi, 1970.

Bhatia, S. C.: *Education and Socio-Cultural Disadvantage,* Xerxes Publications, Delhi, 1982.

Biswas, A. K.: *Science in Pre-Modern India,* D.K. Print World, New Delhi, 2001.

Biswas, Dutt Sunnittee and Singhal R. P.: *The New Educational Pattern in India,* Vikas Publishing House, Delhi, 1916.

Blavatsky, H. P.: *Science: An Introduction,* Concept, London, 1877.

Bose, D. M. and Subbarayappa B. V.: *A Concise History of Science in India,* Oxford, New Delhi, 1971.

Buch, M. B.: *Second Survey of Research in Education 1972-1978,* Society for Educational Research and Development, Baroda, 1979.

Clarke, Adele and Theresa Montini: *Science: Technology and Human Values*, Oxford, New York, 1993.

Denis, Lawton: *Education and Social Justice*, Sage Publications, London, 1977.

Desai, A. R.: *Papers in the Sociology of Education in India*, NCERT, New Delhi, 1990.

Dharni, P. Sinha: *Descriptive-cum-Critical Study of Education System in India*, Administrative Staff College of India, Hyderabad, 1976.

Don, Adams: *Education and Modernisation in Asia*, Addison Wesley Publishing Company, London, 1970.

Duncan, W. A.: *Engendering School Learning: Science, Attitudes and Achievement among Girls and Boys in Botswana*, University of Stockholm, Stockholm, 1989.

Gaur, J. S. and Saraswat, R. K.: *Occupational Literature: An Annotated Bibliography*, NCERT, New Delhi, 1978.

Gill, L. S.: *Science, Technology and Gender*, Polity Press in Association with the Open University, London, 1992.

Goel, S. C.: *Education and Economic Growth in India*, The MacMillan Company of India Ltd., Delhi, 1975.

Gore, M. S.: *Papers in the Sociology of Education in India*, NCERT, New Delhi, 1967.

Gupta S. K.: *Teaching of Physical Science*, Sterling, New Delhi, 1998.

Gupta, S. D. and Sharma D. R.: *Teaching of Science*, Malhotra Bros, Jammu, 2002.

Haragopal, G.: *Administrative Leadership and Rural Development in India*, Light & Life Publishers, New Delhi, 1980.

John, W. Hanson and Cole S. Brem Beck: *Education and the Development of Nations*, Holt, Rinehart and Winston, New York, 1966.

Kochhar, S. K.: *Pivotal Issues in Indian Education*, Sterling, New Delhi, 1981.

Kohli, V. K.: *How to Teach Science,* Vivek Publishers, Ambala City, 2001.

Kotte, Dieter: *Gender Differences in Science Achievement in Ten Countries,* Verlag Peter, Frankfurt, 1992.

Lakshmana, Swamy Mudaliar A.: *Education in India,* Asia Publishing House, Bombay, 1960.

Licuanan, V. S.: *Science and Technology in Asia,* Asian Institute of Management, New Delhi, 1992.

Louis, Malassis: *The Rural World: Education and Development,* Croom Helm, London, 1976.

Lubin, C. R. and Winslow A.: *Science: History,* Duke University Press, London, 1990.

Malhotra, Rajiv and Patel Jay: *History of Indian Science and Technology,* HPH, Mumbai, 2000.

Moussa, Farag: *Science and Technology: Role of Indian Society,* World Intellectual Property Organization, Geneva, 1991.

Mukherji, S. N.: *Administration of Education in India,* Acharya Book Depot, Baroda, 1962.

Niblett, W. R.: *Essential Education,* University of London, London,1955.

Oakes, Jennie: *Lost Talent: The Underparticipation of Women, Minorities and Disabled Persons in Science,* Santa Monica, London, 1990.

Palit, C. B.: *Mahendralal Sircar and National Science,* Ratna Prakashan, Calcutta, 2000.

Prem Kripal: *A Decade of Education in India,* Indian Book Company, Delhi, 1968.

Raghunath, Safaya: *Innovations and Latest Trends in Education,* The Associated Publishers, Ambala Cantt., 1976.

Raju, V. B.: *Commentaries on the Constitution of India,* Eastern Book Company, Lucknow, 1973.

Rudolph and Rudolph: *Education and Politics in India,* Harvard University Press, Cambridge, 1972.

Samba, Murthy K. V.: *Science and Technology in Medieval India*, Pratap, Agra, 1982.

Sandra, Harding: *The "Racial" Economy of Science*, Indiana University Press, New York, 1993.

Segal, Aaron: *Learning by Doing: Science and Technology in the Developing World*, Westview, Boulder, 1987.

Sharma, R. C.: *Modern Science Teaching*, Dhanpat Rai Publishing Co., Delhi, 1981.

Siddiqui, N. N. and Siddiqui N.: *Teaching of Science*, Wilson Publication, New Delhi, 1995.

Stolte-Heiskanen, V.: *Science: Role of Changing India*, ISSC, New Delhi, 1991.

Subbarayappa, B. V.: *Science and Technological Exchanges between India and Soviet Central Asia*, Educational, New Delhi, 1985.

Sunil, Sen Sarma: *Kautilya's Arthasastra in the Light of Modern Science and Technology*, D. K. Printworld(P) Ltd., New Delhi, 2001.

Teresi, Dick: *Lost Discoveries: The Ancient Roots of Modern Science - from the Babylonians to the Maya*, Simon & Schuster, New York, 2002.

Thomas, F. Station: *Dynamics of Adjustment*, MacMillan Co., New York, 1953.

Tiwari, D. D.: *Education at the Cross Roads*, Chugh Publications, Allahabad, 1975.

Turnbull, David and Helen Watson-Verran: *Handbook on Science, Technology and Society*, Sage, Beverly Hills, London, 1994.

Uberoi, Chanchal: *India's Long History of Women in Science*, Physics Today, New Delhi, 1993.

Wardha, B.: *Science and Technology in India*, Centre of Science for Villages, Wardha, 1983.

Zimmerman, J.: *Science in India: Today and Tomorrow*, Praeger Publishers, London, 1983.

# Index


**A**

Administration, 16, 167, 174, 253.
Administrators, 16, 19, 24, 25, 116, 126, 260.
Application, 2, 9, 10, 30, 36, 37, 100, 106, 109, 110, 112, 113, 114, 119, 120, 193, 210, 216, 218, 227, 229, 275, 278, 280, 285, 286, 287, 293.

**B**

Branching Programme, 142.
Broadcast, 173, 174.
Broadmindedness, 34, 39, 114.

**C**

Child Centred Approach, 1, 60, 67, 84, 85, 86.
Classification, 116, 117, 189, 261, 289.
Commission, 17, 30, 32, 106, 111, 126, 130, 191, 195, 196, 199, 202, 205, 208.
Comprehension, 2, 10, 91, 119, 120, 165, 182, 212.
Compulsory Subject, 18, 31, 32.

**D**

Demography, 253.
Demonstrations, 12, 14, 15, 57, 61, 62, 63, 70, 71, 72, 73, 160, 167, 179, 181, 184, 185, 186, 197, 277, 284, 294.
Discovery, 39, 63, 66, 78, 86, 103, 123, 204, 277.
Domain, 117, 118, 119, 121, 123.

**E**

Educational Films, 161, 162.
Educational System, 242, 253.
Educational Technology, 14, 16, 175, 292.
Educational TV, 165, 166, 167, 168.
Evaluation, 2, 15, 16, 17, 25, 49, 107, 115, 116,

119, 120, 136, 138, 139, 149, 166, 168, 169, 203, 215, 246, 249, 252, 288, 292, 293, 294, 295.

Examination, 2, 3, 38, 50, 54, 77, 124, 166, 195, 211, 227, 234, 261, 263, 265.

Experiences, 13, 19, 20, 50, 61, 70, 73, 74, 101, 102, 127, 143, 145, 159, 160, 161, 165, 169, 178, 179, 180, 181, 182, 183, 184, 185, 186, 187, 191, 193, 196, 200, 201, 206, 229, 233, 247, 274, 276, 280, 283, 287, 290, 293.

Experiments, 7, 12, 14, 15, 38, 40, 48, 52, 57, 62, 63, 70, 71, 72, 75, 76, 77, 78, 79, 80, 86, 89, 94, 103, 114, 126, 133, 135, 145, 146, 186, 197, 280, 290, 293, 294.

## F

Feedback, 45, 46, 47, 62, 138, 168, 295.

Field Trip, 40, 176, 179, 185, 186, 188, 189, 190.

Financial, 166, 266.

## G

Generation, 92, 227.

Government, 90, 92, 107, 196, 212, 224.

## H

Human Resources, 196.

## I

Independence, 149, 199.

Interpretation, 41, 63, 79, 113, 136, 208, 213, 228.

Investigation, 3, 7, 8, 10, 64, 77, 78, 94, 103, 201, 212, 220, 266.

## J

Joint Training, 268.

Judgement, 20, 34, 35, 103, 113, 114, 209, 213, 218, 228, 242, 249.

## K

Kothari Commission, 111, 126, 195, 205.

## L

Learn Science, 9, 11, 12, 58, 74, 132.

Lecture Demonstration Method, 66, 75, 83.

Lecture Method, 44, 50, 57, 63, 66, 67, 68, 69, 70, 83, 85, 88.

Linear Programme, 141.

## M

Methodology, 75, 78, 116, 136, 174, 206, 211, 229.

Microteaching, 44, 45, 46, 47.

Multimedia, 15, 61, 67, 89.

## O

Organisation, 44, 45, 55, 113, 117, 120, 122, 197, 203, 204, 217, 259, 267, 269, 275.

## P

Presentation, 4, 70, 73, 74, 161, 167, 168, 169, 172, 275, 277, 280, 284, 285, 286.

Primary, 27, 56, 57, 58, 100, 109, 111, 127, 128, 178, 184, 193, 202, 204, 205, 206, 207, 209, 210, 211, 212, 213, 214, 215, 216, 217, 219, 220, 221, 223, 224, 225, 226, 229, 231, 233, 235, 236, 237, 241, 242, 245, 248, 249, 251, 252, 253, 254, 255, 256, 257, 258, 259, 260, 261, 263, 264, 265, 266, 267, 268, 269, 270, 271, 272.

Programmed Guidance, 89, 144, 149.

Programmed Instruction, 12, 67, 82, 89, 138, 142, 143, 144, 145, 146, 148, 149.

Projectors, 162, 169, 170, 171, 172, 177.

## R

Radio Broadcasts, 173.

Radio Recordings, 61, 169, 181.

Reading Skills, 48, 51, 253.

## S

Satellite, 89, 90.

Science Education, 3, 8, 10, 12, 13, 16, 17, 18, 19, 25, 26, 89, 95, 97, 99, 115, 116, 126, 127, 128, 129, 131, 193, 194, 196, 205, 206, 208, 209.

Scientific Attitude, 8, 10, 19, 20, 21, 23, 24, 30, 35, 39, 59, 100, 103, 104, 108, 110, 113, 114, 130, 131, 193, 276, 293.

Scientific Method, 2, 8, 17, 20, 21, 23, 24, 30, 39, 59, 60, 62, 63, 66, 77, 79, 81, 82, 85, 100,

103, 104, 108, 110, 113, 114, 126, 132, 201, 203.

Secondary Education, 32, 106, 107, 109, 195, 199, 205, 261, 265.

Senior Secondary, 26, 57, 76, 123, 131, 132.

Study of Science, 8, 29, 33, 34, 37, 38, 39, 40, 41, 114, 196.

Synthesis, 119, 120.

**T**

Teacher Centred Approach, 1, 60, 63, 83, 84, 85, 86.

Teacher Training, 91, 222, 242.

Teaching Approach, 43, 61, 63.

Teaching Science, 1, 2, 3, 9, 11, 12, 14, 20, 24, 40, 43, 47, 50, 58, 59, 60, 62, 64, 66, 77, 79, 82, 85, 87, 88, 92, 94, 97, 100, 105, 106, 111, 112, 114, 123, 126, 132, 145, 169, 175, 178, 182, 184, 194, 280, 284, 294.

Teaching Skills, 44, 45, 47, 83.

Teaching Techniques, 65.

Technology, 8, 9, 10, 14, 16, 18, 20, 30, 35, 36, 92, 93, 104, 126, 128, 129, 132, 173, 175, 202, 212, 214, 217, 292.

**U**

Universities, 18, 26, 27, 31, 48, 225.

**V**

Voice Modulation, 172, 173.

**W**

Work Experience, 129.

❑❑❑